Communications
in Computer and Information Science   1320

More information about this series at http://www.springer.com/series/7899

Hong Mei · Weiguo Zhang ·
Wenfei Fan · Zili Zhang ·
Yihua Huang · Jiajun Bu ·
Yang Gao · Li Wang (Eds.)

Big Data

8th CCF Conference, BigData 2020
Chongqing, China, October 22–24, 2020
Revised Selected Papers

 Springer

Editors
Hong Mei
PLA Academy of Military Sciences
Beijing, China

Wenfei Fan
The University of Edinburgh
Edinburgh, UK

Yihua Huang
Nanjing University
Nanjing, China

Yang Gao
Nanjing University
Nanjing, China

Weiguo Zhang
Southwest University
Chongqing, China

Zili Zhang
Southwest University
Chongqing, China

Jiajun Bu
Zhejiang University
Hangzhou, China

Li Wang
Taiyuan University of Technology
Taiyuan, China

ISSN 1865-0929   ISSN 1865-0937   (electronic)
Communications in Computer and Information Science
ISBN 978-981-16-0704-2   ISBN 978-981-16-0705-9   (eBook)
https://doi.org/10.1007/978-981-16-0705-9

This Springer imprint is published by the registered company Springer Nature Singapore Pte Ltd.
The registered company address is: 152 Beach Road, #21-01/04 Gateway East, Singapore 189721, Singapore

Preface

This volume presents the proceedings of the 8th China Computer Federation International Conference on Big Data (CCF BIGDATA 2020), which was held during October 22–24, 2020, in Chongqing, China, with the theme "Big data enabled digital economy".

In the past decade, research on big data has made great progress and many achievements. Now, we enter a new decade of big data, in which we need to continue our efforts to further explore all theoretical and technical aspects of big data, but with more attention to the development trend of the digital economy's demands on big data applications.

CCF BIGDATA 2020 achieved great success in further exploring the theory, techniques, and applications of big data. The submitted papers covered a broad range of research topics on big data, including Big Data Analysis and Applications, Image and Natural Language Big Data, Big Data Intelligent Algorithms, Network Data and Network Science, Big Data and Deep Learning, Big Data Privacy and Security, Big Data Clustering Retrieval, and Big Data Systems. BIGDATA 2020 served as a main forum for researchers and practitioners from academia, industry, and government to share their ideas, research results, and experiences, and to promote their research and technical innovations in the field.

This volume consists of 16 accepted English papers from CCF BIGDATA 2020 and covers most research topics on big data. The papers come from both academia and industry in China, reflecting current research progress and future development trends in big data. All the accepted papers were peer reviewed by three qualified and experienced reviewers. Review forms were designed to help reviewers determine the contributions made by the papers.

CCF BIGDATA 2020 would like to express sincere thanks to CCF, CCF Task Force on Big Data, Southwest University, and our many other sponsors, for their support and sponsorship. We thank all members of our Steering Committee, Program Committee and Organization Committee for their advice and contributions. Also we thank all the reviewers for reviewing papers in such a short time during the reviewing period.

We are especially grateful to the proceedings publisher, Springer, for publishing the proceedings. Moreover, we really appreciate all the keynote speakers, session chairs, authors, conference attendees, and student volunteers, for their participation and contribution.

Program Committee and Organization Committee, the 8th CCF International Conference on Big Data (CCF BIGDATA 2020).

October 2020

Hong Mei
Weiguo Zhang
Wenfei Fan
Zili Zhang
Yihua Huang
Jiajun Bu
Yang Gao

Organization

CCF BigData 2020 was organized by the China Computer Federation and co-organized by the China Computer Federation Task Force on Big Data and Southwest University, China.

Conference Honorary Chairs

Guojie Li   Institute of Computing Technology, Chinese Academy of Sciences, China; Academician of Chinese Academy of Engineering, China

General Chairs

Hong Mei   PLA Academy of Military Science, China; Academician of Chinese Academy of Sciences, China

Weiguo Zhang   Southwest University, China

Wenfei Fan   University of Edinburgh, UK; Shenzhen Institute of Computing Sciences, China

Program Chairs

Zili Zhang   Southwest University, China

Yihua Huang   Nanjing University, China

Jiajun Bu   Zhejiang University, China

Organizing Committee Chairs

Guoqiang Xiao   Southwest University, China

Junhao Wen   Chongqing University, China

Rui Mao   Shenzhen University, China

Publicity Chairs

Qinghua Zhang   Chongqing University of Posts and Telecommunications, China

Chao Gao   Southwest University, China

Laizhong Cui   Shenzhen University, China

Best Paper Award Chairs

Tong Ruan   East China University of Science and Technology,
                           China
Junping Du   Beijing University of Posts and Telecommunications,
                           China

Publication Chairs

Yang Gao   Nanjing University, China
Li Wang   Taiyuan University of Technology, China

Finance Chairs

Li Tao   Southwest University, China
Yuting Guo   Institute of Computing Technology, Chinese Academy
                           of Sciences, China

Sponsorship Chairs

Jianxi Yang   Chongqing Jiaotong University, China
Yilei Lu   Mininglamp Technology, China

Workshop Chairs

Xiaoru Yuan   Peking University, China
Tao Jia   Southwest University, China

Program Committee

Shifei Ding   China University of Mining and Technology, China
Zhiming Ding   Beijing University of Technology, China
Jiajun Bu   Zhejiang University, China
Jianye Yu   Beijing Wuzi University, China
Junqing Yu   Huazhong University of Science and Technology,
                           China
Shuai Ma   Beijing University of Aeronautics and Astronautics,
                           China
Huadong Ma   Beijing University of Posts and Telecommunications,
                           China
Jun Ma   Shandong University, China
Zhixin Ma   Lanzhou University, China
Xuebin Ma   Inner Mongolia University, China
Yuanzhuo Wang   Institute of Computing Technology, Chinese Academy
                           of Sciences, China
Wenjun Wang   Tianjin University, China

Wei Wang	East China Normal University, China
Jin Wang	Suchow University, China
Zhijun Wang	China United Network Communications Group Co. Ltd., China
Can Wang	Zhejiang University, China
Hongzhi Wang	Harbin Institute of Technology, China
Guoren Wang	Beijing Institute of Technology, China
Guoyin Wang	Chongqing University of Posts and Telecommunications, China
Zhiping Wang	Jiangsu United Credit Investigation Co. Ltd., China
Jianmin Wang	Tsinghua University, China
Liang Wang	Institute of Automation, Chinese Academy of Sciences, China
Li Wang	Taiyuan University of Technology, China
Xiaoyang Wang	Fudan University, China
Jianzong Wang	Ping An Technology (Shenzhen) Co., Ltd., China
Hao Wang	Alibaba Group, China
Chongjun Wang	Nanjing University, China
Qi Wang	Northwestern Polytechnical University, China
Tengjiao Wang	Peking University, China
Xin Wang	Fudan University, China
Yijie Wang	National University of Defense Technology, China
Jingyuan Wang	Beijing University of Aeronautics and Astronautics, China
Lei Wang	Institute of Computing Technology, Chinese Academy of Sciences, China
Wenji Mao	Institute of Automation, Chinese Academy of Sciences, China
Rui Mao	Shenzhen University, China
Junhao Wen	Chongqing University, China
Jirong Wen	Renmin University of China, China
Liang Fang	National University of Defense Technology, China
Bo Deng	PLA Academy of Military Science, China
Zhaohong Deng	Jiangnan University, China
Xuanhua Shi	Huazhong University of Science and Technology, China
Yong Shi	University of Chinese Academy of Sciences, China
Chaoqun Zhan	Taobao (China) Software Co., Ltd./Aliyun Intelligent Business Group, China
Yilei Lu	Millward Brown Technology Group, China
Huilin Lu	Wuxi Vocational Institute of Commerce, China
Tun Lu	Fudan University, China
Shaojing Fu	National University of Defense Technology, China
Shuo Bai	Shanghai Danwu Intelligent Technology Co., Ltd., China
Jian Yin	Sun Yat-sen University, China

Jianzhou Feng	Yanshan University, China
Shicong Feng	The refined intelligence, China
Chunxiao Xing	Tsinghua University, China
Xiuzhen Cheng	Shandong University, China
Tuergen Yibulayin	Xinjiang University, China
Xiaofei Zhu	Chongqing University of Technology, China
Wenwu Zhu	Tsinghua University, China
Yangyong Zhu	Fudan University, China
Shunzhi Zhu	Xiamen Institute of Technology, China
Shaojie Qiao	Chengdu University of Information Technology, China
Jiadong Ren	Yanshan University, China
Lei Ren	Beijing University of Aeronautics and Astronautics, China
Yunsheng Hua	The Chinese University of Hong Kong, China
Yang Xiang	Tongji University, China
Shijun Liu	Shandong University, China
Yubao Liu	Sun Yat-sen University, China
Liping Liu	Baidu Online Network Technology (Beijing) Co. Ltd., China
Wei Liu	Dell Technology China R&D Group, China
Chi Liu	Beijing Institute of Technology, China
Qingshan Liu	Nanjing University of Information Technology, China
Jie Liu	Nankai University, China
Xinran Liu	National Computer Network and Information Security Management Center, China
Zheng Liu	SAS Software Research and Development (Beijing) Co. Ltd., China
Yang Liu	Shandong University, China
Hui Liu	Information Center of China Council for the Promotion of International Trade, China
Mengchi Liu	Wuhan University, China
Xuemei Liu	North China University of Water Resources and Electric Power, China
Qi Liu	University of Science and Technology of China, China
Yan Liu	University of Southern California, USA
Jin Tang	Anhui University, China
Jiajie Xu	Suchow University, China
Tong Ruan	East China University of Science and Technology, China
Shaoling Sun	China Mobile Suzhou R&D Center, China
Yunchuan Sun	Beijing Normal University, China
Junchi Yan	Shanghai Jiao Tong University, China
Xiaoyong Du	Renmin University of China, China
Junping Du	Beijing University of Posts and Telecommunications, China

Yuejin Du	Qihoo of Beijing Science and Technology Co Ltd., China
Feifei Li	Alibaba Group, China
Tianrui Li	Southwest Jiaotong University, China
Shijun Li	Wuhan University, China
Dongsheng Li	National University of Defense Technology, China
Guangya Li	Wonders Information Co., Ltd., China
Yufeng Li	Nanjing University, China
Keqiu Li	Tianjin University, China
Qingshan Li	Xidian University, China
Guojie Li	Institute of Computing Technology, Chinese Academy of Sciences, China
Kai Li	University of Science and Technology of China, China
Jianzhong Li	Harbin Institute of Technology, China
Jianxin Li	Beijing University of Aeronautics and Astronautics, China
Ru Li	Inner Mongolia University, China
Xiaoming Li	Peking University, China
Yidong Li	Beijing Jiaotong University, China
Chonggang Li	Jinxin E-Bank Financial Information Service Co. Ltd., China
Ruixuan Li	Huazhong University of Science and Technology, China
Jingyuan Li	Tencent, China
Cuiping Li	Renmin University of China, China
Dongri Yang	China Electronic Information Industry Development Research Institute, China
Ming Yang	Nanjing Normal University, China
Jianxi Yang	Chongqing Jiaotong University, China
Jing Yang	Institute of Computing Technology, Chinese Academy of Sciences, China
Yan Yang	Southwest Jiaotong University, China
Weidong Xiao	National University of Defense Technology, China
Ruliang Xiao	Fujian Normal University, China
Limin Xiao	Beijing University of Aeronautics and Astronautics, China
Guoqiang Xiao	Southwest University, China
Nong Xiao	National University of Defense Technology, China
Gansha Wu	Uisee, China
Jiyi Wu	Zhejiang University Net New 100 Orange Technology Co. Ltd., China
Jun He	Renmin University of China, China
Liwen He	Nanjing University of Posts and Telecommunications, China
Guoliang He	Wuhan University, China
Jieyue He	Southeast University, China

Xiaofei He	Zhejiang University, China
Daojing He	East China Normal University, China
Zhenying He	Fudan University, China
Wei Yu	Wuhan University, China
Guoxian Yu	Shandong University, China
Huawei Shen	Institute of Computing Technology, Chinese Academy of Sciences, China
Shuo Shen	Network Center of Chinese Academy of Sciences, China
Chao Shen	Xi'an Jiaotong University, China
Huaiming Song	Dawning Information Industry Co. Ltd., China
Yulun Song	Unicom Big Data Co. Ltd., China
Yunquan Zhang	Institute of Computing Technology, Chinese Academy of Sciences, China
Yunyong Zhang	China Unicom Research Institute, China
Dong Zhang	Inspur Electronic Information Industry Share Price Co. Ltd., China
Chengqi Zhang	University of Technology Sydney, Australia
Shichao Zhang	Central South University, China
Zili Zhang	Southwest University, China
Jun Zhang	Sun Yat-sen University, China
Liangjie Zhang	Kingdee International Software Group Co. Ltd., China
Mingxin Zhang	Changshu Institute of Technology, China
Xiaodong Zhang	The Ohio State University, USA
Tieying Zhang	Alibaba Silicon Valley Lab, USA
Haijun Zhang	University of Science and Technology Beijing, China
Haitao Zhang	Beijing University of Posts and Telecommunications, China
Qinghua Zhang	Chongqing University of Posts and Telecommunications, China
Jin Zhang	Institute of Computing Technology, Chinese Academy of Sciences, China
Ping Lu	ZTE Corporation, China
Wei Chen	Microsoft Research Asia, China
Shimin Chen	Institute of Computing Technology, Chinese Academy of Sciences, China
Hanhua Chen	Huazhong University of Science and Technology, China
Gang Chen	FiberHome Telecommunication Technologies Co. Ltd., China
Hong Chen	Renmin University of China, China
Shangyi Chen	Baidu Online Network Technology (Beijing) Co. Ltd., China
Xuebin Chen	North China University of Science and Technology, China
Baoquan chen	Peking University, China

Jun Chen	China National Defense Science and Technology Information Center, China
Geng Chen	Nanjing Audit University, China
Enhong Chen	University of Science and Technology of China, China
Jidong Chen	Ant Financial Services Group, China
Ying Chen	Huike Education Technology Group Co. Ltd., China
Wei Chen	Peking University, China
Junming Shao	University of Electronic Science and Technology, China
Yongwei Wu	Tsinghua University, China
Qiguang Miao	Xidian University, China
Youfang Lin	Beijing Jiaotong University, China
Wangqun Lin	Beijing Institute of System Engineering, China
Xuemin Lin	East China Normal University, China
Xiaodong Lin	Rutgers University, USA
Zhonghong Ou	Beijing University of Posts and Telecommunications, China
Jianquan Ouyang	Xiangtan University, China
Shengmei Luo	Cloud Innovation Big Data Technology Co., Ltd., China
Wei Luo	China National Defense Science and Technology Information Center, China
Tongkai Ji	Cloud Computing Industry Technology Innovation and Development Center, Chinese Academy of Sciences, China
Bo Jin	Third Research Institute of the Ministry of Public Security, China
Hai Jin	Huazhong University of Science and Technology, China
Peiquan Jin	University of Science and Technology of China, China
Xin Jin	Central University of Finance and Economics, China
Fengfeng Zhou	Jilin University, China
Yangfan Zhou	Fudan University, China
Xu Zhou	Computer Network Information Center, Chinese Academy of Sciences, China
Yuanchun Zhou	Computer Network Information Center, Chinese Academy of Sciences, China
Xiaofang Zhou	University of Queensland, Australia; Soochow University, China
Tao Zhou	Alibaba Group, China
Aoying Zhou	East China Normal University, China
Bin Zhou	National University of Defense Technology, China
Xia Zhou	Sinopec Petroleum Engineering Geophysics Co. Ltd., China
Zhiwen Yu	Northwestern Polytechnical University, China
Weimin Zheng	Tsinghua University, China

Xiaolong Zheng	Institute of Automation, Chinese Academy of Sciences, China
Yun Zheng	Microsoft China, China
Dongyan Zhao	Peking University, China
Gang Zhao	Academy of Military Systems Engineering, China
Guodong Zhao	Zhongguancun Big Data Industry Alliance, China
Pengpeng Zhao	Soochow University, China
Yong Zhao	University of Electronic Science and Technology, China
Gansen Zhao	South China Normal University, China
Yue Zhao	Unicom Big Data Co. Ltd., China
Xiang Zhao	National University of Defense Technology, China
Bin Hu	Lanzhou University, China
Li Cha	Business-intelligence of Oriental Nations Corporation Ltd., China
Chao Liu	Tian Yan Cha, China
Xiaohui Yu	Shandong University, China
Hengshu Zhu	Baidu Online Network Technology (Beijing) Co. Ltd., China
Jiawei Luo	Hunan University, China
Yongbin Qin	Fudan University, China
Zheng Qin	Hunan University, China
Xiaoru Yuan	Peking University, China
Xiaotong Yuan	Nanjing University of Information Technology, China
Xiaojie Yuan	Nankai University, China
Ye Yuan	Northeastern University, China
Tao Jia	Southwest University, China
Yuanqing Xia	Beijing Institute of Technology, China
Ning Gu	Fudan University, China
Yuhua Qian	Shanxi University, China
Xianghua Xu	Hangzhou Dianzi University, China
Ke Xu	Tsinghua University, China
Sheng Gao	Beijing University of Posts and Telecommunications, China
Yang Gao	Nanjing university, China
Hong Gao	Harbin Institute of Technology, China
Xiaofeng Gao	Shanghai Jiao Tong University, China
Changshui Gao	Ministry of Industry and Information Technology, China
Chao Gao	Southwest University, China
Xinbo Gao	Chongqing University of Posts and Telecommunications, China
Kun Guo	Fuzhou University, China
Yanhong Guo	Dalian University of Technology, China
Bin Guo	Northwestern Polytechnical University, China
Li Tao	Southwest University, China

Zhenghua Xue   Tsinghua University, China
Guirong Xue   Heaven and Earth Smart, China
Hong Mu   Shurui Data Technology Co., Ltd., China
Kai Wei   China Academy of Information and Communications
                         Technology, China
Wei Wei   Xi'an University of Technology, China

Contents

A Short Text Classification Model Based on Cross-Layer Connected Gated Recurrent Unit Capsule Network

Xiaohong Deng, Shiqun Yin$^{(\boxtimes)}$, and Haibo Deng

Faculty of Computer and Information Science, Southwest University,
Chongqing 400715, China

Abstract. Text classification is an important task in natural language processing. In the past few years, some prominent methods have achieved great results in text classification, but there is still a lot of room for development in enhancing the expressive capabilities of text features. In addition, although deeper networks can better extract features, they are easy to produce gradient disappearance or gradient explosion problems. To solve these problems, this paper proposed a hybrid model based on cross-layer connected gated recurrent unit capsule network (CBiGRU_CapsNet). Firstly, the model fused word vectors trained by Word2Vec and GloVe to form the network input layer. In the stage of high-level semantic modeling of text, a cross-layer connected gated recurrent unit has been proposed, which can solve the problem of gradient disappearance or explosion, and strengthen the transfer between features of each layer. Furthermore, capsule network is applied to obtain the rich spatial position information in the deep high-level semantic representation, and the weight value of important features has been increasing through the core dynamic routing algorithm of the capsule network to improve the expression ability of the features, which is then inputted to the softmax layer to complete text classification task. The experimental results on several public datasets shown that CBiGRU_CapsNet outperforms the state-of-the-art models in terms of the accuracy for text classification tasks.

Keywords: BiGRU · Cross-layer · Capsule network · Text classification

1 Introduction

With the rapid development of the mobile Internet and social networks, more and more unstructured short text data has been produced, which usually have strong scientific research value, commercial value, and social value. Therefore, how to filter, manage, and mine these text data quickly and effectively is a research

Supported by the Science & Technology project.

H. Mei et al. (Eds.): BigData 2020, CCIS 1320, pp. 1–17, 2021.
https://doi.org/10.1007/978-981-16-0705-9_1

hotspot in the field of natural language processing, and text classification is the key technology. At present, there are mainly two methods for text classification: 1) Based on traditional machine learning methods; 2) The most popular deep learning methods at this stage.

Machine learning-based text classification methods focus on feature engineering and use different types of machine learning algorithms as classifiers [1]. In the early machine learning text classification tasks, text features were extracted through a series of techniques such as TF-IDF, mutual information, and information entropy. Subsequently, a series of classifiers such as Naive Bayes [2], Support Vector Machine (SVM), KNN [3] and decision tree appeared. GouDjil et al. [4] used the posterior probability of the SVM classifier to select samples for classification, which eased the relationship between text features. Bidi et al. [5] proposed to use a genetic algorithm to improve the process of feature selection in text classification, thereby improving the accuracy of classification. Neethu et al. [6] proposed a text classification model combining the K-Means algorithm and extreme learning machine. Although more complex features are designed to capture more contextual semantics and word order information, they still suffer from data sparseness, which seriously affects the effectiveness of classification.

In recent years, deep learning techniques have been widely used to solve problems related to natural language processing. The basis for applying neural networks to natural language processing is word embedding technology, which maps each word to a dense vector of fixed dimensions. There are two main ways of embedding words: Word2Vec [7] and GloVe [8]. Word2Vec is a word vector model based on prediction, with CBOW and Skip-Gram training modes. GloVe method combines global matrix factorization and local context window method through a bilinear regression model. The most extensive networks in deep learning for text classification are: convolutional neural network [9] and recurrent neural network [10]. The performance of these networks is superior to traditional methods in most tasks, but they are prone to gradient explosion and gradient disappearance problems and there is semantic information that cannot fully understand and encode sentences, so the classification effect needs to be improved.

Since Word2Vec and GloVe can represent the feature information of the text, this paper proposes to combine them and use them as the network input layer to realize the complementary advantages and disadvantages. Then through the cross-layer connected gated recurrent unit to obtain higher-level global semantic features. Next, the spatial location information and local features are extracted through the capsule network, and the weight value of the semantic feature is automatically adjusted through the dynamic routing algorithm to mine more hidden feature information. Finally, the final semantic representation is sent to the softmax layer to realize the classification task. The experiment achieved good results on multiple datasets.

The main works of this paper are as follows:

1) We propose a cross-layer connected gated recurrent unit capsule network (CBiGRU_CapsNet) hybrid classification model to improve the classification accuracy.
2) By combining GloVe and Word2Vec word vector matrix to generate a new vector matrix, which takes into account both local context information and global statistical information, thereby improving the expression ability of word vectors.
3) The cross-layer connected BiGRU network module is used to extract deeper contextual semantic information. In a BiGRU network with a traditional L layer, each layer is connected to the next layer in a feed-forward way. This deep network can solve the problem of disappearing gradients and strengthen the transfer between features of each layer.
4) The capsule network is used to increase the weight of important semantic features and to mine more hidden feature information, so as to improve the accuracy of text classification.

2   Related Work

In recent years, neural network models based on deep learning have achieved great improvement in text classification. Recurrent neural networks (RNN) have attracted much attention in text classification because of their excellent ability to save sequence information for a period of time. But RNN has the phenomenon of gradient disappearance or gradient dispersion. Thereby, some variants of RNN have been put forward, such as LSTM [11], GRU [12]. They overcome the gradient problem of RNN and achieve better results than simple RNN in many different tasks. Nie et al. [13] used the idea of the residual network to stack three layers of bidirectional LSTM, that is, splicing the original sentence representation and the output of the previous LSTM layer, and using it as the input of the next layer of LSTM. This sentence coding method can maximize the learning ability of multi-layer LSTM and prevent the network from improving performance when the number of layers increases to a certain extent. Wei et al. [14] proposed a hybrid sentiment analysis model, which combines BiGRU with Attention mechanism, experimental results show the method can improve the accuracy of sentiment classification.

The combination of RNN and CNN are the most advanced semantic synthesis models for text classification in recent years. Cho et al. [15] proposed a combination model of CNN and RNN for text classification, which achieved good accuracy. Hassan et al. [16] proposed a model that relied on CNN and bidirectional RNN, replacing the pooling layer in CNN with the bidirectional layer, which had high performance. Tang et al. [17] proposed a network model which used CNN or LSTM to generate RNN to adaptively encode sentence semantics and their internal relations. The results showed that using the gated recurrent neural network in this model was significantly better than the standard CNN.

Zhang et al. [19] proposed DSCNN model for text classification, which had great accuracy. The BLSTM-2DCNN model was proposed by Zhou et al. [20], which uses two-dimensional convolution and pooling operations on the sentence. Wang et al. [21] proposed Conv-RNN model, which uses BiGRU to semantically model the sentence, and then performs convolutional pooling operations on it. Liu et al. [22] proposed AC-BiLSTM model for text classification which contains a bidirectional LSTM(BiLSTM), attention mechanism and the convolutional layer, the results show that AC-BiLSTM outperforms other state-of-the-art text classification methods in terms of the classification accuracy.

In recent years, Sabour et al. [23] put forward a method called capsule network for image analysis. They introduce an iterative routing process to decide the credit attribution between nodes from lower and higher layers. Zhao et al. [24] proposed three dynamic routing strategies to mitigate the interference of some noise capsules that may contain redundant information, and applied their method in text classification. The experimental results demonstrate that capsule network outperforms baseline methods in text classification. Kim et al. [25] proposed a static routing algorithm for sentiment classification tasks, which effectively reduced the complexity of routing computation and achieved good results on multiple datasets. Du et al. [26] proposed a capsule model based on BiGRU for sentiment analysis, which utilizes BiGRU and capsule code semantic information together to raise the classification results.

It can be seen that much research has been done in the basic structure of RNN to enhance in its performance and RNN has also achieved outstanding results in text classification. These methods are the basis of CBiGRU_CapsNet and therefore they are elaborated.

3 Proposed Method

In this paper, a hybrid text classification model based on cross-layer connected gated recurrent unit capsule network (CBiGRU_CapsNet) is proposed. The structure of the model is displayed in Fig. 1. It consists of four parts: word vector layer, cross-layer connected gated recurrent unit(CBiGRU) layer, capsule network(CapsNet) layer and softmax classification layer. The key points of our approach are described in detail as follows.

3.1 Word Vector Layer

Text representation is the first step in text classification. The basis of text representation is word embedding technology, which maps each word to a dense vector with a fixed dimension. There are two main methods of word embedding: Word2Vec and GloVe. Word2Vec only considers the relationship between local texts, ignoring the global statistics of words. GloVe integrates global information into word vectors by combining matrix decomposition and local window construction, which can represent global statistical information of words. In order to fully describe the local relations of words and the global information

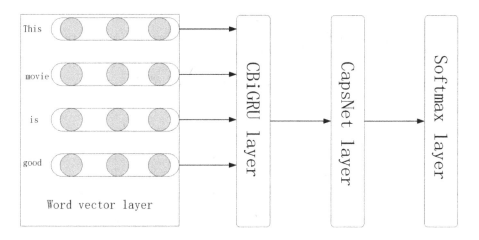

Fig. 1. CBiGRU_CapsNet model.

of texts, we generate a new vector matrix by fusing the vector matrix generated by Word2Vec and GloVe to improve the expression ability of the word vector.

Firstly, Word2Vec is used to vectorize the text. Represent text T in the corpus as a sequence $(t_1, t_2, ..., t_n)$, where n denotes the number of words in text T; t_i denotes the $i - th$ word in T. To convert the text T into a word vector matrix, firstly, the word vector corresponding to ti needs to be searched in the specified word vector space, and if it exists, the corresponding word vector is selected and described by X_i; otherwise, the corresponding vector X_i is set to 0. After all word vectors corresponding to each word is found, each word vector is stacked to form a feature matrix $X \in R^{n \times m}$ (m denotes the dimension of the word vector), and the $i - th$ row of X represents the $i - th$ word in text T. The text T is transformed into a word vector matrix expressed as formula (1).

$$(t_1, t_2, ..., t_n) = [X_1, X_2, ..., X_n]^T \tag{1}$$

Where $X_i = [X_{i1}, X_{i2}, ..., X_{im}]$ represents the m-dimensional Word2vec word vector corresponding to word t_i.

Then GloVe is used to vectorize the text. The process is similar to the word vectorization process of Word2Vec and will not be described here. Each word vector is stacked to form a feature matrix $Y \in R^{n \times m}$ (m denotes the dimension of the word vector), and the $i - th$ row of Y represents the $i - th$ word in text T. The text T is transformed into a word vector matrix expressed as formula (2).

$$(t_1, t_2, ..., t_n) = [Y_1, Y_2, ..., Y_n]^T \tag{2}$$

Where $Y_i = [Y_{i1}, Y_{i2}, ..., Y_{im}]$ represents the m-dimensional Word2Vec word vector corresponding to word t_i.

The new vector matrix E is generated by fusing word2Vec and GloVe as formula (3).

$$E_i = (X_i + Y_i) \in R^m \tag{3}$$

The new matrix generated is shown in Fig. 2.

The final vector matrix E is an n $\times$ m matrix as formula (4), and the $i-th$ row of E indicates the $i-th$ word in text T.

$$E = [E_1, E_2, ..., E_n]^T = (X + Y) \in R^{n \times m} \tag{4}$$

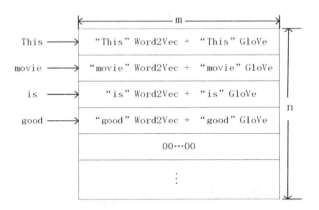

Fig. 2. New word vector matrix.

3.2 CBiGRU Layer

GRU. GRU is a neural network that processes sequence data and consists of update gates and reset gates. The degree of influence of the output hidden layer at the previous moment on the current hidden layer is controlled by the update gate. The larger the value of the update gate, the greater the impact of the output of the hidden layer at the previous moment on the current hidden layer. The degree of the hidden layer information being ignored at the previous moment is controlled by the reset gate. The smaller the reset gate value, the more it is ignored. The mathematical form of GRU shown in Fig. 3 is given.

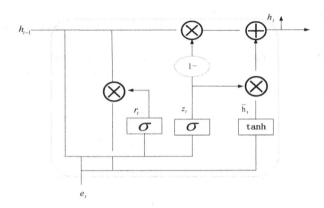

Fig. 3. Illustration of the GRU unit.

The hidden state h_t given input e_t is computed as follows:

$$r_t = \sigma(W_r \cdot [h_{t-1}, e_t]) \tag{5}$$

$$z_t = \sigma(W_z \cdot [h_{t-1}, e_t]) \tag{6}$$

$$\tilde{h}_t = \tanh(W \cdot [r_t * h_{t-1}, e_t]) \tag{7}$$

$$h_t = (1 - z_t) * h_{t-1} + z_t * \tilde{h}_t \tag{8}$$

Where W_r, W_z, W represent the weight matrix, $\sigma()$ and $\tanh()$ are sigmoid activation function and hyperbolic tangent function respectively. $\tilde{h}_t$ represents the candidate state, e_t represents the current input at time t, and the reset gate r_t controls the amount of information that has acted on the candidate state in the past. If the value of r_t is 0, it means that all the previous states are forgotten. The update gate z_t is used to control the amount of past information retained and the amount of new information added.

When GRU models text sequences, the hidden state h_t at each moment t can only forward encode the previous context, regardless of the reverse context. BiGRU make use of two parallel channels, one GRU models the text semantics from the beginning to the end of the sentence, and the other GRU performs the text representation from the end of the sentence to the beginning of the sentence, and then connects the hidden states of the two GRUs as the representation of each moment t. In this way, the output at the current moment is not only related to the previous state, but also to the future state, so that the context can be considered simultaneously. The specific expression is as follows:

$$\overrightarrow{h_t} = GRU\left(\overrightarrow{h_t}, e_t\right) \tag{9}$$

$$\overleftarrow{h_t} = GRU\left(\overleftarrow{h_t}, e_t\right) \tag{10}$$

$$h_t = \left[\overrightarrow{h_t} \oplus \overleftarrow{h_t}\right] \tag{11}$$

Where $\oplus$ refers to the connection function between the two output directions. $\overrightarrow{h}$ stands for forward hidden state output, $\overleftarrow{h}$ stands for backward hidden state output.

Cross-layer Connected Gated Recurrent Unit. Studies have shown that, whether it is a convolutional neural network or a recurrent neural network, deeper depth models can extract more feature information, but stacked depth models can easily lead to the problem of gradient explosions and gradient disappearance, so it is difficult to train the deeper network. Inspired by Densenet [27], we propose a cross-layer connection bidirectional gated recurrent unit (CBiGRU) module to extract deeper contextual semantic information. In a bidirectional GRU network with an L layer, each layer is connected to the next layer in a feed-forward manner. This deep network can solve the problem of gradient disappearance, strengthen the feature transfer between layers, and realize feature reuse. For each bidirectional gated recurrent unit layer, it can directly read the original input feature sequence, so it does not need to transfer all useful information, only need to add information to the network, so that the number of hidden layers in each layer is less. The specific structure of CBiGRU is shown in Fig. 4.

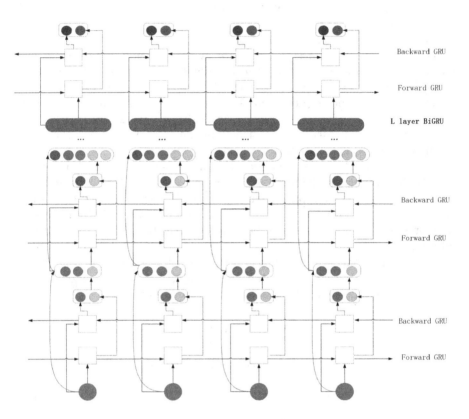

Fig. 4. Cross-layer connected gated recurrent unit.

The one layer input is:
$\{e_1, e_2, ..., e_n\}$
The one layer output is:
$h_1 = \left\{ \left[\overrightarrow{h_1^1}; \overleftarrow{h_1^1}\right], \left[\overrightarrow{h_2^1}; \overleftarrow{h_2^1}\right], ..., \left[\overrightarrow{h_n^1}; \overleftarrow{h_n^1}\right] \right\}$
The two layer input is:
$\left\{ \left[\overrightarrow{h_1^1}; e_1; \overleftarrow{h_1^1}\right], \left[\overrightarrow{h_2^1}; e_2; \overleftarrow{h_2^1}\right], ..., \left[\overrightarrow{h_n^1}; e_n; \overleftarrow{h_n^1}\right] \right\}$
The two layer output is:
$h_2 = \left\{ \left[\overrightarrow{h_1^2}; \overleftarrow{h_1^2}\right], \left[\overrightarrow{h_2^2}; \overleftarrow{h_2^2}\right], ..., \left[\overrightarrow{h_n^2}; \overleftarrow{h_n^2}\right] \right\}$

The three layer input is:

$$\left\{ \left[\overrightarrow{h_1^2}; \overrightarrow{h_1^1}; e_1; \overleftarrow{h_1^1}; \overleftarrow{h_1^2} \right], \left[\overrightarrow{h_2^2}; \overrightarrow{h_2^1}; e_2; \overleftarrow{h_2^1}; \overleftarrow{h_2^2} \right], \ldots, \left[\overrightarrow{h_n^2}; \overrightarrow{h_n^1}; e_n; \overleftarrow{h_n^1}; \overleftarrow{h_n^2} \right] \right\}$$

Through this connection mode, the dimensions of the output of the BiGRU in the previous $L-1$ layer are the same, but the network is a structure with increasing input characteristics. We output the network of the $L-th$ layer $H = \{h_1, h_2, ..., h_n\}$ as the final semantic representation.

3.3 CapsNet Layer

Capsule Network is proposed by Hinton et al. [26,28]. Capsule Network was developed on the basis of CNN. Traditional CNN extracts the underlying local phrase features of the text sequence through convolution kernels. Although the maximum pooling operation reduces the number of network neurons, it causes information loss. The capsule network uses a set of vector neurons (capsules) to replace the scalar neuron nodes in the traditional neural network, which changes the structure of the traditional neural network scalar and scalar connection. The information carried by each capsule in each network layer has increased from one dimension to multiple dimensions, retaining the local order and semantic representation of the words in the text. The lower layer capsule transmits the calculation results saved in this layer to the upper layer capsule through the Dynamic Routing mechanism [25], so as to reduce the loss of information while extracting local feature phrases, and can increase the weight of important features and hidden features. Thereby improving the accuracy of classification.

The input of capsule network is h_t that is the output from previous layer CBiGRU. The operations are obtained as follow:

$$\hat{u}_{j|i} = W_{ij}h_t \tag{12}$$

$$s_{out} = \sum_{i=1}^{m} c_{ij}\hat{u}_{j|i} \tag{13}$$

$$c_{ij} = \frac{\exp(b_{ij})}{\sum_k \exp(b_{ik})} \tag{14}$$

W_{ij} represents the weight matrix that determines the correlation between the input layer and the output layer. c_{ij} is the coupling coefficient which is updated iteratively dynamic routing algorithm. The sum of coupling coefficient between the capsules of input layer and output layer is 1. It is computed by softmax with initialized b_{ij} to 0.

The non-linear activation function squash is adopted to normalize the output vector shown as follow:

$$v_{out} = \frac{\|s_{out}\|^2}{1 + \|s_{out}\|^2} \frac{s_{out}}{\|s_{out}\|} \tag{15}$$

The dynamic routing algorithm is summarized in Table 1.

Table 1. The process of dynamic routing algorithm.

Algorithm 1 Dynamic Routing Algorithm

Procedure Routing(u_i, r, l)

For all capsule i in layer l and capsule j in $l + 1$

Initial: $\hat{u}_{j|i} \leftarrow W_{ij}u_i, b_{ij} \leftarrow 0$

for r interations do

      for all capsule i in layer l:$c_{ij} \leftarrow soft\max(b_{ij})$

      for all capsule i in layer l and capsule j in $l + 1$:$s_{out} \leftarrow \sum_i c_{ij}\hat{u}_{j|i}$

      for all capsule j in $l + 1$:$v_{out} = squash(s_{out})$

      for all capsule j in $l + 1$:$b_{ij} \leftarrow b_{ij} + \hat{u}_{j|i} \cdot v_{out}$

end

Return v_{out}

3.4   Softmax Layer

The semantic matrix finally extracted by the capsule network is input into the dropout layer to prevent overfitting problems. The feature matrix extracted by capsule network is input into a dropout layer to prevent the over-fitting problem. During the training process, some neurons which are selected randomly do not work, but they are still retained for the next input sample. The other neurons participate in the process of computation and connection. The vector matrix is input into a full connection layer for dimension reduction. Finally, the softmax activation function is used to calculate the probability distribution of this category. The specific calculation formula is as follows:

$$p_y = soft\max(W_v V' + b) \tag{16}$$

Where W_v is a fully connected layer weight matrix optimized by a stochastic gradient descent algorithm, and b is a bias term. V' represents the feature vector of the final sentence feature vector V after being processed by the dropout layer.

Currently, the cross entropy is a commonly used loss function to evaluate the classification performance of the models. It is often better than the classification error rate or the mean square error. In our work, Adam optimizer is chosen to optimize the loss function of the network. The model parameters are fine-tuned by Adam optimizer which has been shown as an effective and efficient back propagation algorithm. The cross entropy as the loss function can reduce the risk of a gradient disappearance during the process of stochastic gradient descent. The loss function can be denoted as follows:

$$\text{loss} = -\sum_{i=1}^{r}\sum_{j=1}^{c} y \log y' + \lambda\|\theta\|^2 \tag{17}$$

Where r is the training dataset size, c is the number of classes, y' is the predicted class, and y is the actual class. $\lambda\|\theta\|^2$ is the regular terms.

4 Experiment

4.1 Dataset

The statistical results of the experimental dataset are shown in Table 2. In Table 2, c is the number of categories classified by the dataset, l is the average sentence length, m is the maximum sentence length, train/dev/test is the number of samples in the training set/validation set/test set, and CV means no standard training set/test set is divided, and the 10-fold cross-validation is adopted in the experiment of this paper to evaluate the accuracy of the algorithm.

Table 2. Corpora statistics.

Data	c	m	l	train	dev	test
MR	2	59	21	10662	–	CV
Subj	2	65	23	10000	–	CV
TREC	6	33	10	5452	–	CV
SST-1	5	51	18	8544	1101	2210
SST-2	2	51	19	6920	872	1821

The relevant data sets in the experiment are described as follows:

MR (Movie Review): Dataset of Movie reviews, the task is to detect positive/negative reviews;

Subj (Subjectivity): The subjective dataset. The task is to classify sentences as subjective or objective.

TREC: Problem classification dataset. This task involves identifying the type of a problem and dividing it into six problem types, namely debris, description, entity, person, location, and value.

SST-1 (Stanford Sentiment Treebank): Stanford University Sentiment Database is an extension of MR data set by Socher et al. The purpose is to classify reviews into fine-grained labels, that is, very negative, negative, neutral, positive, and very positive.

SST-2: Same as the SST-1 data set, but the neutral comments have been deleted, and only contain the secondary classification labels, that is, negative and positive.

4.2 Experiment Settings

In this work, the datasets have all been divided into training and test sets, and the SST also includes the divided validation set. For other datasets that do not contain a standard validation set, this work randomly selects 10% of the training dataset as the validation set.

Our experiments use accuracy as the evaluate metric to measure the overall classification performance. Word2Vec and GloVe word vectors as pre-trained word embeddings. The size of these embeddings is 300. The number of layers about the cross-layer connected is 10. The last layer has a hidden layer unit of 100. The dimension of capsule is set to 128, and the iteration number of dynamic routing is 3. To minimize marginal loss, the stochastic gradient descent method is adopted as the model training optimization algorithm, and the learning rate is 0.01. The mini-batch size is 64, and the parameter of dropout is 0.5.

4.3 Baseline Methods

To evaluate the effectiveness of our model for sentiment classification, our model is compared to the following baseline models, including some simple neural network methods such as CNN-static, CNN-non-static, CNN-multichannel [9], and BiLSTM, and some combined methods such as BLSTM-2DCNN [20], C-LSTM [18], DSCNN [19], Conv-RNN [21], Capsule-A, Capsule-B [24], AC-BiLSTM [22] and DC-BiGRU_CNN [29].

4.4 Result

Overall Comparison. The experiments results on all datasets are shown in Table 3. The best results are shown in **boldface**. For some baselines, we directly use the results given in the original paper, and we reproduce the results for the rest. We adopt the accuracy (higher is better) metric to evaluate the performance of the model on each dataset. It can be noticed that the accuracies for some methods are missing because the corresponding results cannot be found anywhere in relevant literature.

Table 3. Experimental results of classification accuracy (%).

Dataset	MR	Subj	TREC	SST-1	SST-2
CNN-static	81.5	93.4	93.6	48.0	87.2
CNN-non-static	81.0	93.0	92.8	45.5	86.8
CNN-multichannel	81.1	93.2	92.2	47.4	88.1
BiLSTM	80.3	92.3	95.5	49.1	87.5
C-LSTM	–	–	94.6	49.2	87.8
BLSTM-2DCNN	82.3	94.0	96.1	**52.4**	–
Conv-RNN	82.0	94.1	–	51.7	88.9
DSCNN	81.5	93.2	95.4	49.7	89.1
Capsule-A	81.3	93.3	91.8	–	86.4
Capsule-B	82.3	93.8	92.8	–	86.8
AC-BiLSTM	83.2	94.0	–	48.9	88.3
DC-BiGRU_CNN	82.6	94.1	95.0	51.2	88.5
CBiGRU_CapsNet	**83.4**	**94.5**	**96.3**	51.9	**89.2**

From Table 3, compared with the CNN model, RNN model, and other advanced hybrid models in the five datasets tested, CBiGRU_CapsNet achieves better results than other methods, especially the accuracy rates of MR, Subj, TREC and SST-2 were 83.4%, 94.5%, 96.3% and 89.2%, respectively. In view of the above discussion it can be concluded that the overall performance of CBi-GRU_CapsNet is better than that of the most state of-the-art methods in terms of the classification accuracy.

Effect of Each Component of CBiGRU_CapsNet. CBiGRU_CapsNet contains three components, namely, the word vector layer, cross-layer connected Gated Recurrent Unit (CBiGRU), the Capsule Network layer. For CBi-GRU_CapsNet, it should be proven that all components are useful for the final results. In this section, a set of experiments are to investigate the effect of each component on the performance of CBiGRU_CapsNet. Therefore, this paper designed three comparative models: a) CBiGRU_CapsNet only uses Word2Vec to vectorize the word matrix as the input layer of the network, and the other components are unchanged No-GloVe-CBiGRU_CapsNet model; b) In the cross-layer connection module, the capsule network is not used, and other components do not unchanged CBiGRU model; c) BiGRU_CapsNet model without cross-layer connection components. The results are shown in Table 4.

Table 4. Effect of each component on the performance of CBiGRU_CapsNet (%).

Dataset	MR	Subj	TREC	SST-1	SST-2
BiGRU_CapsNet	81.9	93.8	95.1	50.5	87.9
CBiGRU	82.9	94.5	95.8	51.5	88.5
No-GloVe-CBiGRU_CapsNet	82.5	94.1	95.0	51.2	88.3
CBiGRU_CapsNet	**83.4**	**94.5**	**96.3**	**51.9**	**89.2**

According to the results shown in Table 4, No-GloVe-CBiGRU_CapsNet, CBi-GRU, and BiGRU_CapsNet have lower accuracy than the CBiGRU_CapsNet in the test set, which proves that our integrated Word2Vec/GloVe word embedding, cross-layer connected BiGRU, capsule network and other components is very effective.

Effect of Sentence Length on CBiGRU_CapsNet. In general, when we train the model, we need to specify a maximum sentence length value. In order to study the influence of the selection of the maximum sentence length on the accuracy of the classification task, we selected the MR dataset for experiments. The sentence length distribution in this paper is shown in Fig. 5.

The experimental results are shown in Fig. 6. For the MR dataset, the maximum sentence length is 35, and the effect is the best. If the maximum length value is small, the semantic feature information will be lost when the text is intercepted. If the maximum length value is larger, the effect of padding on the accuracy of the model will also increase.

Effect of BiGRU Network Depth on CBiGRU_CapsNet. In this paper, the MR dataset is selected to study the effect of the number of cross-layer connected layers on the model performance. The experimental results are shown in Fig. 7. The results show that under certain conditions, the performance of the model improves with the increase of the number of network layers, but if the number of layers is too deep, the accuracy of the model does not improve, but declines. It can be seen from the experiment that the model has the best effect when the number of layers is about 10.

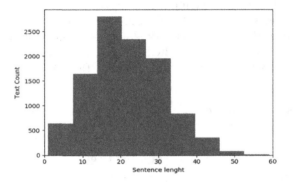

Fig. 5. Sentence length distribution of MR.

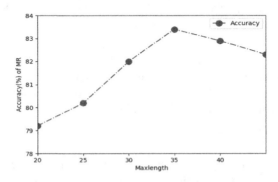

Fig. 6. Impact of maximum length of sentence on accuracy.

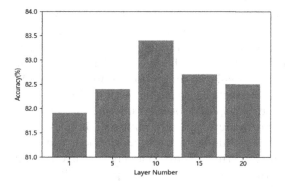

Fig. 7. Impact of layer number of cross-layer network on model's accuracy.

Effect of Dynamic Routing Iteration of CBiGRU_CapsNet. In this paper, MR dataset is selected to study the impact of the number of iterations of the dynamic routing algorithm on the capsule network on the performance of model. The experimental results are shown in Fig. 8. The results show that under certain conditions, the performance of the model improves as the number of iterations of the dynamic routing algorithm increases, but if there are too many iterations, the accuracy of the model does not improve, but declines. The experiment shows that the dynamic routing algorithm has the best effect when the number of iteration is 3.

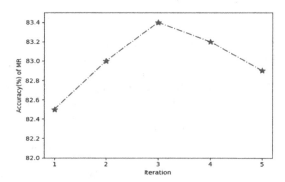

Fig. 8. The Influence of the iteration of dynamic routing on the model accuracy.

5   Conclusion

For the task of short text classification, this paper proposes a hybrid model based on cross-layer connected gated recurrent unit capsule network (CBi-GRU_CapsNet). Firstly, a new vector matrix generated by fusing GloVe and Word2Vec was used as the input layer of the network, it can not only describe the local information between words, but also notice the global information of the text, thus improving the expression ability of word vectors. Then, the global semantic modeling is carried out through the cross-layer connected gated recurrent unit to make full use of all the information learned from the lower level, and to establish the cross-layer connection between the low-level features and high-level features to enrich the semantic features. Finally, the capsule network is used to extract richer spatial location information, and the weight value of the semantic feature is automatically adjusted through the dynamic routing algorithm to mine more hidden feature information, which improves the classification accuracy. Experiments are conducted on five benchmark datasets to evaluate the performance of our proposed approach. The results indicate that CBiGRU_CapsNet can understand semantics more accurately and enhance the performance of the text classification. Comparisons with some state-of-the-art baseline methods, it demonstrates that the new method is more effective in most cases.

In the future work, we will explore how to join attention mechanism, so that our model can further improve classification accuracy.

Acknowledgment. This work is supported by the Science & Technology project (41008114, 41011215, and 41014117).

References

1. Joachims, T.: Text categorization with support vector machines: learning with many relevant features. In: Nédellec, C., Rouveirol, C. (eds.) ECML 1998. LNCS, vol. 1398, pp. 137–142. Springer, Heidelberg (1998). https://doi.org/10.1007/BFb0026683
2. Chen, Z., Shi, G., Wang, X.: Text classification based on Naive Bayes algorithm with feature selection. Int. J. Inf. **15**(10), 4255–4260 (2012)
3. Vries, A., Mamoulis, N., Nes, N., et al.: Efficient KNN search on vertically decomposed data. In: Proceedings of SIGMOD 2002, pp. 322–333 (2002)
4. Goudjil, M., Koudil, M., Bedda, M., Ghoggali, N.: A novel active learning method using SVM for text classification. Int. J. Autom. Comput. **15**(3), 290–298 (2016). https://doi.org/10.1007/s11633-015-0912-z
5. Bidi, N., Elberrichi, Z.: Feature selection for text classification using genetic algorithms. In: Proceedings of ICMIC 2016, pp. 806–810 (2016)
6. Neethu, K., Jyothis, T., Dev, J.: Research on Web text classification algorithm based on improved CNN and SVM. In: Proceedings of ICCT 2017, pp. 1958–1961 (2017)
7. Mikolov, T.: Distributed representations of words and phrases and their compositionality. In: Advances in Neural Information Processing Systems, vol. 26, pp. 1–5 (2013)
8. PNennington, J., Socher, R., et al.: GloVe: global vectors for word representation. In: Proceedings of EMNLP 2016, pp. 1532–1543 (2016)
9. Kim, Y.: Convolutional neural networks for sentence classification. In: Proceedings of EMNLP 2014, pp. 1746–1751 (2014)
10. Zhang, Y., Chen, G., Yu, D., et al.: Highway long short-term memory RNNS for distant speech recognition. In: Proceedings of IEEE 2016, pp. 5755–5759 (2016)

11. Hochreiter, S., Schmidhuber, J.: Long short-term memory. Neural Comput. **9**(08), 1735–1780 (1997)
12. Chung, J., Gulcehre, C., et al.: Empirical evaluation of gated recurrent neural networks on sequence modeling. CoRR (2014)
13. Nie, Y., Bansal, M.: Shortcut-stacked sentence encoders for multi-domain inference. In: Proceedings of EMNLP 2017, pp. 41–45 (2017)
14. Wei, W., Yu, S., Qing, Q., et al.: Text sentiment orientation analysis based on multi-channel CNN and bidirectional GRU with attention mechanism. IEEE Access **8**, 134964–134975 (2020)
15. Xiao, Y., Cho, K.: Efficient character-level document classification by combining convolution and recurrent layers. CoRR (2016)
16. Hassan, A., Mahmood, A.: Efficient deep learning model for text classification based on recurrent and convolutional layers. In: Proceedings of ICMLA 2017, pp. 1108–1113 (2017)
17. Tang, D., Qin, B., Liu, T.: Document modeling with gated recurrent neural network for sentiment classification. In: Proceeding of EMNLP 2015, pp. 1422–1432 (2015)
18. Zhou, C., Sun, C., Liu, Z., et al.: A C-LSTM neural network for text classification. Comput. Sci. **1**(04), 39–44 (2016)
19. Zhang, R., Lee, H., Radev, D.: Dependency sensitive convolutional neural networks for modeling sentences and documents. In: Proceeding of NAACL 2016, pp. 1512–1521 (2016)
20. Zhou, P., Qi, Z., Zheng, S., et al.: Text classification improved by integrating bidirectional LSTM with two-dimensional max pooling. In: Proceeding of COLING 2016, pp. 3485–3495 (2016)
21. Wang, C., Jiang, F., Yang, H.: A hybrid framework for text modeling with convolutional RNN. In: Proceeding of SIGKDD 2017, pp. 2061–2069 (2017)
22. Liu, G., Guo, J., et al.: Bidirectional LSTM with attention mechanism and convolutional layer for text classification. Neurocomputing **337**(14), 325–338 (2019)
23. Sabour, S., Frosst, N., et al.: Dynamic routing between capsules. In: NIPS, pp. 3856–3866 (2017)
24. Zhao, W., Ye, J., Yang, M., et al.: Investigating capsule networks with dynamic routing for text classification. In: Proceedings of EMNLP 2018, pp. 3110–3119 (2018)
25. Kim, J., Jang, S., et al.: Text classification using capsules. Neurocomputing **376**, 214–221 (2020)
26. Du, Y., Zhao, X., He, M., et al.: A novel capsule based hybrid neural network for sentiment classification. IEEE Access **7**, 39321–39328 (2019)
27. Huang, G., Liu, Z., et al.: Densely Connected Convolutional Networks. In: Proceedings of IEEE CVPR 2017, pp. 2261–2269 (2017)
28. Hinton, E., Sabour, S., et al.: Matrix capsule with EM routing. In: Proceedings of ICLR 2018, pp. 1–29 (2018)
29. Zheng, C., et al.: DC-BiGRU_CNN model for short text classification. Comput. Sci. **46**(11), 186–192 (2019)

Image Compressed Sensing Using Neural Architecture Search

Nan Zhang, Jianzong Wang$^{(\boxtimes)}$, Xiaoyang Qu, and Jing Xiao

Ping An Technology (Shenzhen) Co., Ltd., Shenzhen, China

Abstract. Deep learning methods have been widely applied in image compressed sensing (CS) recently, which achieve a significant improvement to traditional reconstruction algorithms in both running speed and reconstruction quality. However, it is a time-consuming procedure even for an expert to efficiently design a high-performance network for image CS because of various combination of different kernel size and filter number in each layer. In this paper, a novel image CS framework named NAS-CSNet is presented by leveraging virtues from neural architecture search (NAS) technique. The NAS-CSNet includes a sampling network, an initial reconstruction network and a NAS-based reconstruction network, which are optimized jointly. In particular, the reconstruction network is automatically designed by searching from the search space without trials and errors by experts. Extensive experimental results demonstrate that our proposed method achieves the competitive performance compared with the state-of-the-art deep learning methods and numerically promotes the reconstruction accuracy considerably, showing the effectiveness of the proposed NAS-CSNet and the promise to further use of NAS in the CS field.

Keywords: Compressed sensing · Deep learning · Neural Architecture Search · Image reconstruction · Cuckoo search

1 Introduction

Compressed sensing (CS) [1], which breaks through the limitation of sampling frequency in the Nyquist-Shannon sampling theorem, is regarded as a new paradigm of sampling theory. Thus, it becomes one of the hottest topics in the signal processing field and attracts a lot of researchers as well in the last decade. Mathematically, the main purpose of CS is to find a certain solution of an underdetermined linear system, which is described as:

$$\mathbf{y} = \mathbf{A}\mathbf{x} + \mathbf{w} \tag{1}$$

where $\mathbf{x} \in \mathbb{R}^n$ is the signal, $\mathbf{y} \in \mathbb{R}^m$ is the measurement after sampling $\mathbf{x}$ with the sensing matrix $\mathbf{A} \in \mathbb{R}^{m \times n}(m \ll n)$, and $\mathbf{w}$ is the possible noise introduced

This paper is supported by National Key Research and Development Program of China under grant No. 2018YFB1003500, No. 2018YFB0204400 and No. 2017YFB1401202.

H. Mei et al. (Eds.): BigData 2020, CCIS 1320, pp. 18–30, 2021.
https://doi.org/10.1007/978-981-16-0705-9_2

by the sampling process. Candès and Tao [2] have proved in theory that this problem can be solved with a very high probability when it satisfies two fundamental requirements: (1) the signal $\mathbf{x}$ is sparse or could be sparsified with help of a certain transform, which means the number of nonzero elements s is much less than n; (2) the sensing matrix $\mathbf{A}$ satisfies the so-called restricted isometry property (RIP) [3]. As we know that image has a lot of redundant information and could be well represented sparsely, thus image can be naturally compressed and reconstructed.

In the past few years, methods aimed to this CS reconstruction problem could be divided into several catagories: (a) greedy algorithms (e.g. CaSaMP [4]); (b) convex optimization algorithms (e.g. GPSR [5]); (c) Bayesian methods (e.g. BCS [6]); (d) deep learning methods (e.g. DCS [7], BSSR-LSTM [8]). On account of the superior performance both on reconstruction accuracy and computational complexity, deep learning based methods nowadays have been increasingly utilized in this field, especially in image compressed sensing.

Recently, with the tremendous development of Convolutional Neural Network (CNN), a few deep learning based methods [9–13] start to pay great attention to image compressed sensing. In [9], Kulkarni et al. proposed a novel CNN-based non-iterative architecture, ReconNet, in which the intermediate reconstruction is fed into an off-the-shelf denoiser to obtain the final reconstructed image. In [11], a scalable network called LAPRAN is proposed, which could simultaneously generate multiple images with different resolutions for high-fidelity, flexible, and fast CS image reconstruction. Moreover, CSNet [10], proposed by Shi et al., could avoid blocking artifact caused by the above methods via learning an end-to-end mapping between measurements and the original image. Soon afterwards they improved this network and proposed dubbed CSNet [12] and dubbed SCSNet [13]. Specifically, dubbed CSNet could learn the sensing matrix adaptively from the training images besides learn an end-to-end mapping, and dubbed SCSNet could achieve scalable sampling and scalable reconstruction with only one model, providing both coarse and fine granular scalability.

Although deep neural networks have been widely applied in CS problem and achieved state-of-the-art performance, it is still a challenge to design high-performance architectures efficiently. In fact, it is a time-consuming procedure even for an expert who has profound domain knowledge to design appropriate networks because of various combination of different kernel size and filter number in each layer. To address this dilemma, the *Neural Architecture Search* (NAS) technique draws our attention, which explores automating the procedure of designing neural architectures without trials and errors of experts while achieves state-of-the-art performance in lots of applications such as such as object detection [14], image classification [15] and language modeling [16], et al.

In essence, the NAS technique consists of three main components, i.e. search space, search strategy and performance estimation strategy. The search space consists of all possible candidate architectures which might be discovered. The search space can defined simple chain-like neural networks [17] and more complex multi-branch neural networks with skip connections [15,18]. For reducing the size of the search space, the search space is designed by repeating the same structure called cell or block [19]. Search strategy is used to search for the best architecture from the search space, such as evolutionary methods [20], reinforcement learning (RL) [21], and gradient-based methods [22]. RL methods could achieve state-of-the-art performance [23], while usually requiring thousands of GPU days to find networks with high performance. Gradient-based optimization could efficiently identify the promising architecture from a complex search space, along with reducing the time cost by orders of magnitude. Evolutionary methods is also a efficient method which has been successfully applied in NAS problem [15,24], in which network representation and mutation are cores and need to be well designed.

The major contributions of this work are in the following aspects:

- Proposing a novel end-to-end image CS framework called NAS-CSNet, which could automatically design reconstruction network to reconstruct the original image without manually designing via the NAS technique.
- Employing the cuckoo algorithm as the search strategy to search high-performance reconstruction networks.
- Conducting extensive experiments to verify the effectiveness of NAS-CSNet on the image CS task. Experimental results illustrate that, compared with manually designed architectures, the proposed approach achieves the state-of-the-art reconstruction quality.

2   Model Design

Our NAS-CSNet implements block-based image CS via three procedures, i.e., a sampling network, an initial reconstruction network and a NAS-based reconstruction network, as illustrated in Fig. 1. The sampling network is used to learn the sensing matrix and acquire CS measurements simultaneously. The initial reconstruction network, which is a linear operation, is applied to obtain the initial reconstruction from measurements. And the NAS-based reconstruction network is a nonlinear operation to further enhance the reconstruction quality which employs the NAS technique to automatically search for the best neural architecture for image reconstruction.

2.1   Sampling Network

Generally speaking, Gaussian and Bernoulli matrices are usually utilized as the sensing matrix in conventional image CS. With the development of deep learning, convolution or fully-connected layers could be considered for sampling as well. In this work, a convolution layer with n_B filters of kernel size $B \times B$ are served to imitate block-based compressed sampling process. Note that there is no bias and no activation in convolution layers, $n_B = \lfloor \frac{m}{n} \times B^2 \rfloor$ where B denotes block size. Such learned sensing matrix, which could capture more image structural information than general matrices, will be continuously updated during the training.

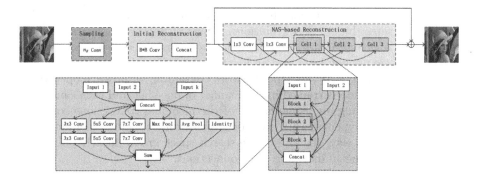

Fig. 1. Diagram of the NAS-CSNet. Solid lines indicate present components in every architecture while dashed lines indicate optional components.

2.2   Initial Reconstruction Network

Given the CS measurements, the initial reconstructed image with the original size could be obtained by a convolution layer with corresponding kernel size and stride similar to the compressed sampling process. Since image blocks will transform into a series of $1 \times 1 \times n_B$ vectors after the sampling network and the size of corresponding initial reconstructed blocks is $B \times B$, B^2 filters of size $1 \times 1 \times n_B$ are used to obtain initial reconstructed blocks. Then the initial reconstructed image will be obtained by concatenating these blocks, which provides a chance to make the best of both intra-block and interblock information to optimize the entire image rather than independent image blocks.

2.3   NAS-Based Reconstruction Network

Since the initial reconstruction network is relatively simple, the image constructed from it has the same size but extremely poor quality compared with the original image. Therefore, a deeper reconstruction network is essential to further refine the reconstructed image. Although currently proposed reconstruction methods based on deep learning are powerful and flexible, designing an architecture with excellent performance still is a challenge, which requires a large amount of trials by experts with rich knowledge and experience.

In this paper, the NAS technique is applied to automatically design a reconstruction network. A well-performance architecture can be automatically searched as final reconstruction network from the search space which is a complex and immense hyper-network. To be specific, the search space is designed on three different levels, i.e. the network, the cell and the block, as shown in Fig. 1.

The **network** is constructed by the several identical stacked cells. Each cell will be connected with up to two subsequent cells. The **cell** is divided into a fixed number of choice blocks. The inputs to a given choice block might come from outputs of two most recent previous cells and outputs of previous choice blocks within the same cell. The **block** includes all candidate operations, and each one could select up to two operations from six optional operations, including a pair of depthwise separable 3×3 convolutions, a pair of depthwise separable 5×5 convolutions, a pair of depthwise separable 7×7 convolutions, a max pooling layer, an average pooling layer and an identity.

It is worth noting that the hyper-network containing all possible operations and all different inputs is trained in the manner of one-shot and each candidate architecture containing any combination of the incoming connections could be inherited from this hyper-network by zeroing out or removing the other incoming connections. In this way, the size of the search space grows linearly only with respect to the number of incoming skip-connections.

2.4   Loss Function

Since the proposed NAS-CSNet follows an end-to-end learning structure, inputs and labels are identical as the ground truth image. Then the mean square error is adopted as the loss function in our model for training the initial reconstruction network and the NAS-based reconstruction network, as follows:

$$L_{init}(\theta, \phi) = \frac{1}{2N} \sum_{i=1}^{N} \|I(S(x_i; \theta); \phi) - x_i\|_2^2 \tag{2}$$

$$L_{NAS}(\theta, \phi) = \frac{1}{2N} \sum_{i=1}^{N} \|R(S(x_i; \theta); \phi) - x_i\|_2^2 \tag{3}$$

where $S(x_i; \theta)$ are CS measurements, $I(S(x_i; \theta); \phi)$ and $R(S(x_i; \theta); \phi)$ are the initial reconstructed output and the final output with respect to the image x_i, respectively. θ and ϕ are parameters of the sampling network and the reconstruction network. N denotes the total number of training samples, x_i represents the i-th sample.

3   Implement Details

Basically, our main goal is to obtain the best reconstruction architecture from the search space, there are three main procedures to achieve: (i) train a single hyper-network which could predict the performance of all possible architectures; (ii) select the candidate architecture with the best validation performance; (iii) retrain the final selected architecture.

3.1   Training Hyper-Network

In general NAS approaches, each possible architecture in the search space is enumerated to train from scratch for a fixed number of epochs and then evaluate on a validation set, which are incredibly resource-hungry. Sharing weights between architectures is a promising direction to amortize the cost of training. Instead of training thousands of separate architectures from scratch, we could train a single but huge hyper-network containing all possible operations, which could emulate any specific architecture in the search space. Every architecture directly inherits the weight parameters of the trained hyper-network, which significantly reduces the computational complexity. At evaluation time, each architecture in the search space could be simulated by enabling or disabling specific combination of incoming connections, and zeroing out or removing some of operations. Actually, removing any component even unimportant ones from the hyper-network may lead to reduced quality of reconstruction and reduced correlation of accuracy between the hyper-network and stand-alone models. To avoid this issue, path dropout is added at training time. The rate of dropout is a linear schedule.

3.2 Best-Accuracy Model Selection and Retraining

After finishing the training, the search strategy that could automatically search the best architecture in the search space will be applied. In this work, we employ the cuckoo search algorithm for a quick architecture selection with the best accuracy.

Cuckoo search is one of the latest nature-inspired metaheuristic algorithms developed by Yang et al. [25] based on the brood parasitism of cuckoos. Recent theoretical researches indicate that cuckoo search has global convergence [26]. There are following three ideal assumptions in cuckoo search: (1) each cuckoo will lay one egg at a time and dumps it in a randomly selected nest; (2) the best nest with high-quality eggs will carried over to the next generation; (3) the number of available host nests is fixed, and the egg of cuckoo will be discovered by the host bird with a probability $p_a \in (0, 1)$. In our problem, the cuckoos and the eggs represent the candidate architectures. The aim is to use the new and potentially better architecture (cuckoo) to replace the not-so-good architecture in the nest.

Explore by Lévy Flight: The goal of exploration is to discover individual diverse. Lévy flight is used to realize this goal. The Lévy flight essentially provides a random walk while the random step length is drawn from a Lévy distribution. One of the most efficient and yet straightforward ways of applying Lévy flights is to use the so-called Mantegna algorithm [27]. In Mantegna's algorithm, the step length S is calculated by $S = u/(|v|^{\frac{1}{\beta}})$, where u and v are obtained from normal distributions as $u \sim N(0, \sigma_u^2)$, $v \sim N(0, \sigma_v^2)$, β is a parameter between the interval $[1, 2]$ and here it is considered as 1.5, $\sigma_u = \left\{ \frac{\Gamma(1+\beta)sin(\pi\beta/2)}{\Gamma[(1+\beta)/2]\beta 2^{(\beta-1)/2}} \right\}^{1/\beta}$, and Γ is the standard Gamma function.

Generate New Cuckoos by Mutation: Lévy flights is used to generate new cuckoos(new models with different architectures), which enhance the diversity of the *population* and guarantee the reasonable balance between local and global evolution. To construct new model, there are two main transformations, one is a control of enabling or disabling incoming connections and the other is a simple and random modification of the operation in the block, are applied here. It is worth noting that only one of these two transformations will be selected in each loop. The implementation algorithm of mutation is in Algorithm 1, where $nest_i^t$ represents the i-th nest of t-th generation and $nest_i^{t+1}$ is the next generation at current position. $\texttt{Transformation}(nest_i^t)$ is a function that could apply transformations to $nest_i^t$.

Algorithm 1. Mutation

Input: Current nest $nest_i^t$
Output: New cuckoo $nest_i^{t+1}$
1: Let $n = 0$, $S = \text{Lévy}(\beta)$
2: **while** $n < S$ or $n < \texttt{len}(nest_i^t)$ **do**
3:     $nest_i^{t+1} = \texttt{Transformation}(nest_i^t)$
4:     $n = n + 1$
5: **end while**
6: **return** $nest_i^{t+1}$

Build New Nest: The cuckoo eggs that lodged with nests will be discovered with a probability p_a, which causes these nests will be abandoned. To keep the number of the *population*, the cuckoo eggs that have not been found will be remained, and new nests will be built near the abandoned nest by transformations if they are discovered, as follows:

$$nest_i^{t+1} = \begin{cases} nest_i^t & \text{if} \quad rand < p_a \\ \texttt{Transformation}(nest_i^t) & \text{if} \quad rand > p_a \end{cases} \tag{4}$$

where $rand$ is a random number between 0 and 1.

The cuckoo search we used can be summarized in Algorithm 2. *Population* set is initialized by trained architectures that selected from the search space randomly with the same probability. Each candidate model is called the nest or the cuckoo, and the head leader architecture in each generation will be recorded in *history*. Here, the architecture with the highest validation fitness will be selected from *sample* as the head leader, where the fitness is defined as Peak Signal to Noise Ratio(PSNR) in this paper. Notice that the *sample* is a small subset of nests that randomly sampled from the *population* with the same probability. Repeat above procedures until the number of architectures in the *history* is greater than C. And the best architecture will be chosen from the *history* by the fitness. Finally, this best-performing architecture will be retrained from scratch for more specific weights.

4   Experiment

4.1   Training Setting

We conduct experiments on public BSDS500 dataset which has 200 training images, 200 test images and 100 validation images. Note that training and test

---

Algorithm 2. Cuckoo search algorithm

Parameter: discovering probability p_a, the number of *population P*, the number of *sample Sa*, and the number of *history C*

1: Generate an initial *population* of P host nests (architectures)
2: **while** $|history| < C$ **do**
3:    Generate a new cuckoo i randomly by mutation and calculate its fitness F_i
4:    Select a nest j randomly and calculate its fitness F_j
5:    **if** $F_i > F_j$ **then**
6:        Replace nest j by new cuckoo i
7:    **end if**
8:    Discovery alien eggs and build new nests
9:    Sample Sa nests randomly from *population* as a candidate set *sample* $(Sa < P)$
10:    Sort and find the current best one $nest_{\text{best}}$ in *sample*
11:    Add $nest_{\text{best}}$ to *history*
12: **end while**
13: **return** The best nest in *history*

---

images (totally 400 images) are employed for training and validation images are still employed for validation. As well as CSNet [12], training images will be augmented by eight data augmentation methods, like flip, rotation, etc. Then these samples will be transformed into grayscale images and finally divided into sub-images of size 96×96 with stride of 57. Finally, we randomly select 89600 sub-images for network training.

Before training our model, some hyper-parameters, such as parameters in the NAS model, the training procedure and the cuckoo search algorithm, etc., are required to be determined. In the search space, the number of cells, choice blocks and filters are set to 4, 3 and 32, respectively. In the training, we set the batch size is 64, the optimizer is Adam, the epoch is 100, the dropout rate is 0.5 and the learning rate is linearly decayed from 0.1 to 0.001. As for the cuckoo search algorithm part, the discovering probability is set to 0.5, and the number of models in *population* set, *sample* set and *history* are set to 100, 25 and 20000, respectively.

4.2   Experimental Results

Reconstruction Accuracy Comparison: Some deep learning methods, including SDA [28], ReconNet [9], ISTA-Net [29] and CSNet [12], are utilized for comparison. Table 1 shows the average PSNR on Set11 with 5 sampling ratios. As shown in Table 1, NAS-CSNet obtains competitive performance with these existing deep learning methods. Compared to the best deep learning, i.e. CSNet[+], NAS-CSNet can improve average PSNR roughly by 0.16 dB, 0.25 dB, 0.54 dB, 0.34 dB, 0.39 dB, respectively, under different sampling ratios. A portion of visual results are illustrated in Fig. 2 for a more intuitional comparison. From partial details of enlargements, the reconstruction of ReconNet is obviously blurry, and ISTA-Net[+], CSNet[+] and NAS-CSNet achieve much clearer results.

The NAS-CSNet can recover finer and sharper details than the other methods that the root in the enlargement is easy to identify, while the results of the other methods are blurry. Besides, the reconstruction of ReconNet and ISTA-Net$^+$ have significant blocking artifacts, while both CSNet$^+$ and NAS-CSNet utilize interblock information to avoid this issue. Compared to the reconstruction of CSNet$^+$, ours has a slightly better effect and shows more details.

Running Speed Comparisons: Traditional image CS methods take roughly several seconds (even several minutes) to reconstruct a 256×256 image, while deep learning based methods take less than one second on CPU or 0.05 s on GPU. The average running times on CPU/GPU of recent deep learning methods for reconstructing a 256×256 image are listed in Table 1. Specifically, NAS-CSNet run faster than other deep learning methods except SDA, but NAS-CSNet significantly outperforms SDA on reconstruction quality.

Table 1. Average PSNR (dB) performance comparisons on Set11 with different CS ratios.

Algorithm	Sampling ratio					Time/s	
	0.5	0.4	0.3	0.1	0.01	CPU	GPU
SDA	28.95	27.79	26.63	22.65	17.29	–	**0.0032**
ReconNet	31.50	30.58	28.74	24.28	17.27	0.5258	0.016
ISTA-Net	37.43	35.56	32.91	25.80	17.30	0.923	0.039
ISTA-Net$^+$	38.07	36.06	33.82	26.64	17.34	1.375	0.047
CSNet*	37.51	36.10	33.86	28.10	20.94	0.2941	0.0155
CSNet$^+$	38.52	36.48	34.30	28.37	21.03	0.9024	0.0257
NAS-CSNet	**38.68**	**36.73**	**34.84**	**28.71**	**21.42**	**0.2528**	0.0132

4.3   Discussion

Basically, the impact of hyper-parameters and search strategy on the final reconstruction results is significant.

Search Space: The search space plays an important role for the search result. Compared with simple chain-like structure, the complex multi-branches structure can improve the performance a little but increase the search space exponentially. The cell/block structure can save computation cost by stacking identical cells. Besides, sharing weights of hyper-network can further accelerate the training procedure. As a result, the cell/block structure is adopted.

Hyper-parameters: The number of cells and filters are two important hyper-parameters for NAS-CSNet, which are evaluated by following experiments. We explore the number of parameters, the number of floating-point operations

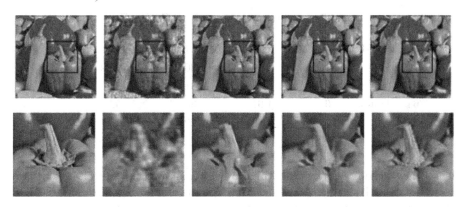

Fig. 2. The visual quality comparison of respective deep learning based CS recovery algorithms on Peppers from Set5 in the case of the sampling ratio is 0.10. The first column are original image and its partial enlargement, the second to the fifth column are reconstructed images using ReconNet (PSNR = 24.19 dB), ISTA-Net$^+$ (PSNR = 27.36 dB), CSNet$^+$ (PSNR = 27.85 dB) and our NAS-CSNet (PSNR = 28.33 dB), respectively.

(FLOPs), PSNR, and training time for different numbers of cells. Table 2 shows that when the number of filters is set to a fixed value of 16, the model size and training time increase with the double increase of the number of cells, but PSNR decreases. The optimal number of cells is 4. Then we fix the number of cells to 4 and change the number of filters. Experimental results in Table 3 show that with double growth of the number of filters, the parameter size and FLOPs continue to increase but PSNR will increase at first but then decrease. The optimal number of filters is 32.

Table 2. The impact of different cell numbers on NAS-CSNet (BDS100, ratio = 0.2).

Cells	Path	Param (10^6)	FLOPs (10^9)	PSNR (dB)	Time (GPUh)
4	all-on	0.51	0.19	30.98	2.72
8	all-on	0.61	0.39	30.57	5.42
16	all-on	0.81	0.73	28.45	11.17
32	all-on	1.20	1.58	25.32	31.83

Search Strategy: To evaluate the performance for cuckoo search as search strategy, we compared cuckoo search with random search. With the same number of sampled architectures, the cuckoo search can find more high-performance architectures compared with the random search strategies. For example, the number of architecture whose fitness greater than 30 from 20000 architectures searched by the cuckoo search and random search is 7673 and 2565, respectively.

Table 3. The impact of different filter numbers on NAS-CSNet (BDS100, ratio $= 0.2$).

Filters	Path	Param (10^6)	FLOPs (10^9)	PSNR (dB)	Time (GPUh)
16	all-on	0.51	0.19	30.98	2.72
32	all-on	0.69	0.57	31.20	3.58
64	all-on	1.36	1.91	30.82	5.5
128	all-on	3.79	6.89	29.65	10.47

Meanwhile, cuckoo search can find high-precision architectures much faster than the random search strategies. Besides, the experiments shows that random search method can reach a high accuracy but the performance is not stable.

5   Conclusion

In this paper, a novel image compressed sensing framework called NAS-CSNet based on the neural architecture search technique is proposed to deal with the time-consuming problem of designing the network structure. Specifically, the NAS-CSNet includes a sampling network, an initial reconstruction network and a NAS-based reconstruction network, where the NAS-based reconstruction network is the best-accuracy architecture automatically searched out from the search space without any trials and possible errors caused by human experts. In addition, the cuckoo search algorithm is employed to seek for the high-performance reconstruction network for the first time. Extensive experimental results demonstrate that the NAS-CSNet outperforms other deep learning based methods on public benchmark datasets. Moreover, the results also illustrate the efficacy and potential of the NAS technique in other fields.

References

1. Donoho, D.L.: Compressed sensing. IEEE Trans. Inf. Theory **52**(4), 1289–1306 (2006)
2. Candès, E.J., Tao, T.: Near-optimal signal recovery from random projections: universal encoding strategies? IEEE Trans. Inf. Theory **12**(52), 5406–5425 (2006)
3. Candès, E.J.: The restricted isometry property and its implications for compressed sensing. Comptes Rendus Mathematique **9**(346), 589–592 (2008)
4. Needell, D., Tropp, J.: CoSaMP: iterative signal recovery from incomplete and inaccurate samples. Appl. Comput. Harmonic Anal. **3**(26), 301–321 (2009)
5. Figueiredo, M.A., Nowak, R.D., Wright, S.J.: Gradient projection for sparse reconstruction: application to compressed sensing and other inverse problems. IEEE J. Sel. Topics Signal Process. **1**(4), 586–597 (2007)
6. Ji, S., Xue, Y., Carin, L.: Bayesian compressive sensing. IEEE Trans. Signal Process. **6**(56), 2346–2356 (2008)
7. Wu, Y., Rosca, M., Lillicrap, T.: Deep compressed sensing. In: Proceedings of the 36th International Conference on Machine Learning, pp. 6850–6860 (2019)

8. Lyu, C., Liu, Z., Yu, L.: Block-sparsity recovery via recurrent neural network. Signal Process. **154**, 129–135 (2019)
9. Kulkarni, K., Lohit, S., Turaga, P., Kerviche, R., Ashok, A.: ReconNet: non-iterative reconstruction of images from compressively sensed measurements. In: Proceedings of the IEEE Conference on Computer Vision and Pattern Recognition, pp. 449–458 (2016)
10. Shi, W., Jiang, F., Zhang, S., Zhao, D.: Deep networks for compressed image sensing. In: 2017 IEEE International Conference on Multimedia and Expo (ICME), pp. 877–882. IEEE (2017)
11. Xu, K., Zhang, Z., Ren, F.: LAPRAN: a scalable laplacian pyramid reconstructive adversarial network for flexible compressive sensing reconstruction. In: Ferrari, V., Hebert, M., Sminchisescu, C., Weiss, Y. (eds.) ECCV 2018. LNCS, vol. 11214, pp. 491–507. Springer, Cham (2018). https://doi.org/10.1007/978-3-030-01249-6_30
12. Shi, W., Jiang, F., Liu, S., Zhao, D.: Image compressed sensing using convolutional neural network. IEEE Trans. Image Process. **29**, 375–388 (2019)
13. Shi, W., Jiang, F., Liu, S., Zhao, D.: Scalable convolutional neural network for image compressed sensing. In: Proceedings of the IEEE Conference on Computer Vision and Pattern Recognition, pp. 12290–12299 (2019)
14. Ghiasi, G., Lin, T.-Y., Le, Q.V.: NAS-FPN: learning scalable feature pyramid architecture for object detection. In: Proceedings of the IEEE Conference on Computer Vision and Pattern Recognition, pp. 7036–7045 (2019)
15. Real, E., Aggarwal, A., Huang, Y., Le, Q.V.: Regularized evolution for image classifier architecture search. In: Proceedings of the AAAI Conference on Artificial Intelligence, pp. 4780–4789 (2019)
16. Luong, M.T., Dohan, D., Yu, A.W., Le, Q.V., Zoph, B., Vasudevan, V.: Exploring neural architecture search for language tasks. In: 6th International Conference on Learning Representations, ICLR, May 2018
17. Cai, H., Chen, T., Zhang, W., Yu, Y., Wang, J.: Efficient architecture search by network transformation. In: Proceedings of the AAAI Conference on Artificial Intelligence, pp. 2787–2794 (2018)
18. Elsken, T., Metzen, J.H., Hutter, F.: Efficient multi-objective neural architecture search via Lamarckian evolution. In: 7th International Conference on Learning Representations, ICLR, May 2019
19. Zoph, B., Vasudevan, V., Shlens, J., Le, Q.V.: Learning transferable architectures for scalable image recognition. In: Proceedings of the IEEE Conference on Computer Vision and Pattern Recognition, pp. 8697–8710 (2018)
20. Liu, H., Karen Simonyan, C.F., Vinyals, O., Kavukcuoglu, K.: Hierarchical representations for efficient architecture search. In: 6th International Conference on Learning Representations, ICLR, May 2018
21. Baker, B., Gupta, O., Naik, N., Raskar, R.: Designing neural network architectures using reinforcement learning. In: 5th International Conference on Learning Representations, ICLR, April 2017
22. Liu, H., Simonyan, K., Yang, Y.: DARTS: differentiable architecture search. In: 7th International Conference on Learning Representations, ICLR, May 2019
23. Real, E., et al.: Large-scale evolution of image classifiers. In: Proceedings of the 34th International Conference on Machine Learning, ICML, August 2017, pp. 2902–2911
24. Lu, Z., et al.: NSGA-Net: neural architecture search using multi-objective genetic algorithm. In: Proceedings of the Genetic and Evolutionary Computation Conference, pp. 419–427. ACM (2019)
25. Yang, X.-S., Deb, S.: Cuckoo search via Lévy flights. In: 2009 World Congress on Nature & Biologically Inspired Computing (NaBIC), pp. 210–214. IEEE (2009)

26. Wang, F., He, X., Wang, Y., Yang, S.: Markov model and convergence analysis based on cuckoo search algorithm. Comput. Eng. **38**(11), 180–185 (2012)
27. Mantegna, R.N.: Fast, accurate algorithm for numerical simulation of Lévy stable stochastic processes. Phys. Rev. E **49**(5), 4677–4683 (1994)
28. Mousavi, A., Patel, A.B., Baraniuk, R.G.: A deep learning approach to structured signal recovery. In: 2015 53rd Annual Allerton Conference on Communication, Control, and Computing (Allerton), pp. 1336–1343. IEEE (2015)
29. Zhang, J., Ghanem, B.: ISTA-Net: interpretable optimization-inspired deep network for image compressive sensing. In: Proceedings of the IEEE Conference on Computer Vision and Pattern Recognition, pp. 1828–1837 (2018)

Discovery of Sparse Formula Based on Elastic Network Method and Its Application in Identification of Turbulent Boundary Layer Wall Function

Tao Ye[1(✉)], Hui Zhang[1(✉)], and Xinguang Wang[2(✉)]

[1] The Institute of Computer Science and Technology,
Southwest University of Science and Technology, Mianyang 621010, China
zhanghui@swust.edu.cn
[2] The Computational Aerodynamics Institute, CARDC,
State Key Laboratory of Aerodynamics, Mianyang 621000, China
wangxinguang@cardc.cn

Abstract. Extracting governing equations from data is a central challenge in diverse areas of science and engineering. Where data are abundant whereas models often remain elusive, as in climate science, neuroscience, ecology, finance, and epidemiology. A sparse representation algorithm based on elastic network optimization is proposed, which combines the advantages of least squares and Lasso to identify the function form directly from the data. The method designs the candidate function terms according to background knowledge, and uses the elastic network optimization algorithm to identify the unknown coefficients of each term with the model. The approach maintains a balance between accuracy and model complexity, avoiding overfitting. The wall function method is a commonly used for dealing with the turbulent boundary layer, which can accelerate convergence rate compared to the near-wall turbulence modelling with reasonable accuracy. The proposed algorithm is used to derive the wall function from the turbulent wall data. Experimental results show that the proposed algorithm is superior to LASSO and least squares in obtaining the model, and faster than numerical calculation.

Keywords: Machine learning · Regression algorithm · Sparse identification · Wall function

1 Introduction

Equations, especially Partial Differential Equations (PDE), play a prominent role in many disciplines to describe the governing physical laws underlying a given system. Traditionally, PDEs are derived mathematically or physically based on some basic principles, for example, from Boltzmann equations could derive to Navier-Stokes equations [1]. However, the mechanisms behind many complex systems in modern applications are still generally unclear, and the governing equations of these systems are commonly obtained by empirical formula, such as the problems in climate science, neuroscience, finance, biological science, etc. With the rapid development of computational power and

© Springer Nature Singapore Pte Ltd. 2021
H. Mei et al. (Eds.): BigData 2020, CCIS 1320, pp. 31–44, 2021.
https://doi.org/10.1007/978-981-16-0705-9_3

data storage in the last decade, huge quantities of data could be easily collected, stored and processed. The vast quantity of data offers new opportunities for discovering physical laws from data, and provides a new idea for deep understanding of complex systems. At present, the main methods of modeling from data can be divided into traditional machine learning algorithms and deep neural network models.

Many deep neural networks emphasize expression ability and prediction accuracy. In most cases, they can meet the observation data and make accurate predictions, but these networks cannot reveal potential equation models [2]. Xu [3] developed a data-driven method based on deep learning, called DL-PDE, for discovering the control partial differential equations of the underlying physical processes. The DL-PDE method uses neural network for deep learning and sparse identification for data-driven PDE discovery. Raissi [4] proposed Physical Information Neural Network (PINN) to solve the positive and inverse problems of PDE. PINN improved the accuracy of results, by adding a PDE constraint term in the loss function. PDE in this algorithm has a known structure, only needs to learn the coefficients of PDE terms from the data, which limits the application of this algorithm. In this regard, Raissi [5] modified the PINN by introducing two neural networks for approximating the unknown solution. Although this modification enables PINN to solve the problem of unknown PDE structure, the neural network approximation of unknown PDE still taken as a black box, lacking interpretability. Long [6] proposed a numeric-symbolic hybrid deep network on the basis of previous work [7]. The method combines the numerical approximation of convolution differential operator with symbolic multilayer neural network, revealing the hidden PDE model by learning the dynamics of complex systems. Gradient disappearance and gradient explosion may occur in the process of solving neural networks. Haber [8] proposed a new deep neural network structure inspired by the system of Ordinary Differential Equations (ODE) to improve the stability of forward propagation. The network structure can overcome the problems of gradient disappearance and gradient explosion.

Traditional machine learning algorithms mainly adopt sparse identification technology, which shows great potential in discovering the control equations of various systems. Brunton [9] introduced machine learning algorithms from the perspective of fluid mechanics, and summarized the applications of machine learning in flow modeling, optimization, and control. The purpose of using sparse identification is to identify a small number of terms that constitute the governing equation from a predefined large candidate library, and obtain a reduced model. Brunton [10] proposed a Sparse Identification of Nonlinear Dynamics (SINDy) framework to discover the governing equations of dynamic systems. Combining the sparse technique and symbolic regression to avoid overfitting, and proposed the Sequential Threshold Least Square (STLS) method to obtain regression coefficients. However, the framework faces huge challenges in finding basic functions of spatio-temporal data or high-dimensional measurements. Rudy [11] solved this limitation using the Sequential Threshold Ridge (STR) regression algorithm, and produced a PDE Functional Identification of Nonlinear Dynamics (PDE-FIND) framework. Rudy [12] then proposed a data-driven method to discover parametric partial differential equations. The Sequential Group Threshold Ridge (SGTR) used in the method is superior to the group lasso algorithm in identifying the minimum terms and parameter dependencies in the equations. Schaeffer [13] proposed an algorithm that uses

L1 regularized least squares to select the correct features for controlling PDE. Zhang [14] applied the data-driven method proposed by Rudy to flow molecular simulation and derived the flow equation. Mangan [15] developed an implicit scheme for non-linear dynamic sparse identification. Schaeffer and McCalla [16] select sparse models through integral terms. Boninsegna [17]. uses the sparsity technique to discover stochastic dynamic equations. Schaeffer [18] found high-dimensional dynamics from limited data. Berg and Nyström [19] identify PDE in complex data sets. Chang and Zhang [20] identify the physical process by combining data-driven and data assimilation. Loiseau [21] combines nonlinear sparse identification with dimension reduction techniques to identify nonlinear reduced-order models of fluid. Li [22] proposed a "structured" method for solving a continuous PDE model with constant and spatial variation.

Many algorithms are used to obtain unknown coefficients for sparse solution. The least squares method will identify all function terms, so that all coefficients are non-zero values, which may lead to overfitting. LASSO algorithm has strong ability to compress coefficients, but its stability is weakened. Ridge regression has high stability but not good in compressing coefficients. Elastic network algorithm is a mixture of LASSO and ridge regression, combining the advantages of both algorithms. It is good in coefficients compression and ensure stability. We propose an algorithm based on elastic network optimization, called Sequentially Threshold Elastic Net (STEN), and use it to identify wall functions.

2 Wall Function

The turbulent boundary layer can be divided into three sublayers: viscous sublayer, buffer layer, log-law layer. The physical laws are varied due to the importance of molecular viscosity and turbulent viscosity. In engineering, wall functions are often used to express the physical laws of speed u^+, temperature T, with the dimensionless wall distance y^+. Normally, the wall function is a two-layer segmented model containing only sublayer and log-law layer for simplifying, which will lead to the discontinuity of velocity and temperature. In order to overcome this problem, Spalding [23] suggested a unified wall function which is valid across the turbulence boundary layer, which is given by

$$y^+ = u^+ + e^{-\kappa B}\left[e^{\kappa u^+} - \sum_{k=0}^{3} \frac{\left(ku^+\right)^n}{n!} \right] \tag{1}$$

where constants κ and B are generally taken as 0.4 and 5.5 respectively.

Nichols and Nelson [24] introduce the compressibility and heat transfer effects into (1) with the outer velocity form of White and Christoph [25], which can be expressed as:

$$y^+ = u^+ + y^+_{white} - e^{-\kappa B} \sum_{k=0}^{3} \frac{\left(ku^+\right)^n}{n!} \tag{2}$$

Where:

$$y^+_{white} = \exp((\kappa/\sqrt{\Gamma})\{\sin^{-1}[(2\Gamma u^+ - \beta)/Q] - \phi\}) \times e^{-\kappa B}$$

In the compressible boundary layer, the temperature wall function is proposed by Crocco-Busemann formula [24]:

$$T/T_w = 1 + \beta u^+ - \Gamma(u^+)^2 \tag{3}$$

Where:

$$\Gamma = \frac{ru_\tau^2}{2C_pT_w}, \beta = \frac{q_w\mu_w}{\rho_wT_wk_wu_\tau}, Q = \sqrt{\beta^2 + 4\Gamma}$$

$$\phi = \sin^{-1}\left(\frac{-\beta}{Q}\right), r = 0.72^{1/3}, Cp = 1004.5$$

The parameter Γ represents compressibility effects, the parameter β represents heat transfer effects, τ_w and q_w represent the wall shear stress and the wall heat transfer respectively. μ_w is the wall molecular viscosity, ρ_w is wall density, T_w is wall temperature, k_w is the wall turbulent kinetic energy, and u_τ is the friction velocity.

3 Sparse Identification Algorithm

The physical system has a large amount of input data and response data, need to determine an equation from these data to represent the quantitative relationship between them. The regression problem in machine learning is used to determine the quantitative relationship between two or more variables. Therefore, the equation found from data can be regarded as a regression problem. This paper uses sparse identification algorithm to deal with regression problems. Collecting the data and designing the nonlinear function candidate library of the regression model according to the data, then using the sparse identification method to find the equations from data. Sparse identification can avoid overfitting as well as reduce the complexity of the function by compressing coefficients [10].

Collecting distance y^+ and velocity u^+ data by solving wall function. Assuming that both of them have the following rule:

$$y^+ = f\left(u^+\right) \tag{4}$$

The function $f\left(u^+\right)$ represents the dynamic constraint of speed on distance. To determine the function f, collected data and arranged it into two matrices:

$$u^+ = \left[u_1^+ \, u_2^+ \, u_3^+ \, \ldots \, u_n^+\right]^T$$
$$y^+ = \left[y_1^+ \, y_2^+ \, y_3^+ \, \ldots \, y_n^+\right]^T \tag{5}$$

We construct a library $\Theta\left(u^+\right)$ consisting of candidate nonlinear functions of the columns of u^+. For example, $\Theta\left(u^+\right)$ may consist of constant, polynomial, and trigonometric terms:

$$\Theta\left(u^+\right) = \left[u^{+^P}, \ldots, log\left(u^+\right), \sqrt{u^+}, \sin\left(u^+\right) \ldots\right] \tag{6}$$

Where $p \in N$. Each column of $\Theta(u^+)$ represents a candidate function for the right-hand side of (4). We may set up a sparse regression problem to determine the sparse vectors of coefficients $\xi = [\xi_1, \xi_2, \xi_3, \ldots, \xi_n]$ for determining which nonlinearities are active:

$$y^+ = \Theta(u^+)\xi \tag{7}$$

The candidate function library $\Theta(u^+)$ and the collected data are brought into the (7):

$$
\begin{bmatrix} y_1^+ \\ y_2^+ \\ y_3^+ \\ \cdots \\ y_n^+ \end{bmatrix}
=
\begin{bmatrix}
(u_1^+)^p & \log(u_1^+) & e^{u_1^+} & \ldots & \sin(u_1^+) & \ldots \\
(u_2^+)^p & \log(u_2^+) & e^{u_2^+} & \ldots & \sin(u_2^+) & \ldots \\
(u_3^+)^p & \log(u_3^+) & e^{u_3^+} & \ldots & \sin(u_3^+) & \ldots \\
& & & \cdots & & \\
(u_n^+)^p & \log(u_n^+) & e^{u_n^+} & \ldots & \sin(u_n^+) & \ldots
\end{bmatrix}
\begin{bmatrix} \xi_1 \\ \xi_2 \\ \xi_3 \\ \cdots \\ \xi_n \end{bmatrix}
\tag{8}
$$

The function form is obtained by solving the coefficient matrix ξ. Each ξ_k of ξ is a sparse vector of coefficients determining which terms are active in the right-hand side of (4).

The ξ in (7) can be obtained by using the least square method, but the least square method tends to fit all function terms, resulting in the coefficient matrix being non-zero and overfitting. Therefore, sparse identification is used to identify the function terms and coefficient estimates required by the system. Sparse identification is used to solve (7), so that most coefficients are zero to achieve the purpose of a simplified model. In the problem of finding equations from data, LASSO, STLS, and ridge regression are used to solve coefficient [10, 13]. Combining the advantages of these three algorithms, this paper proposes a new sparse solution algorithm called Sequential Threshold Elastic Network (STEN) to get the coefficient vector.

3.1 Sparse Algorithm

Traditional linear regression is prone to overfitting. There are two methods could be used to avoid overfitting, one is reducing the number of features, the other is regularization. Reducing the number of features will cause incomplete information and large model errors. Therefore, regularization is the mainstream method to avoid overfitting. Robert [26] proposed the Least Absolute Shrinkage and Selection Operator (LASSO) algorithm, which is a compression estimation method based on the idea of shrinking the variable set. By adding a penalty function to linear regression, the coefficients of variables are compressed and some regression coefficients become zero to achieve the purpose of variable selection. The penalty function is the cost function of the L1 regularization term, so the algorithm is also called the L1 regularization algorithm [16]:

$$\xi = \underset{\xi'}{\arg\min} \left\| \Theta\xi' - y^+ \right\|_2 + \lambda \left\| \xi' \right\|_1 \tag{9}$$

where λ is called the regularization parameter. LASSO makes the coefficients with smaller absolute values directly become zero. LASSO is particularly suitable for the reduction of the number of parameters, selecting parameters, and be used to estimate the regression model of sparse parameters.

Another method is STLS proposed by Brunton [10]. We start with a least-squares solution for ξ and then threshold all coefficients that are smaller than the cutoff value ψ. Once the indices of the remaining non-zero coefficients are identified, we obtain another least-squares solution for ξ onto the remaining indices. These new coefficients are thresholding by using ψ again, and the procedure is continued until the non-zero coefficients converge. This algorithm is computationally efficient, and it rapidly converges to a sparse solution in a small number of iterations.

Based on the above two algorithms, we designed the STEN. It combines the advantages of LASSO, ridge regression and STEN. Both sparsity and stability are ensured. LASSO regression adds L1 regularization term on the basis of loss function, objective function is Eq. 9. The ridge regression adds an L2 regularization term on the basis of the loss function, and the objective function is:

$$\xi = \operatorname*{argmin}_{\xi'} \left\| \Theta \xi' - y^+ \right\|_2 + \lambda \left\| \xi' \right\|_2^2 \tag{10}$$

The elastic network [27] is a linear regression model that uses L1 and L2 norms as a priori regular terms. It is a combined model of LASSO and ridge regression. Its objective function is:

$$\xi = \operatorname*{argmin}_{\xi'} \left\| \Theta \xi' - y^+ \right\|_2 + \lambda_1 \left\| \xi' \right\|_1 + \lambda_2 \left\| \xi' \right\|_2^2 \tag{11}$$

Where λ_1 and λ_2 regularize the weights of the penalty functions L1 and L2, respectively. In (11), λ_1 and λ_2 control the size of sparsity by applying more weight to the penalty term, which can reduce the coefficient and avoid overfitting. A regularized weight parameter α and mixed parameter λ can be used to modify (11):

$$\xi = \operatorname*{argmin}_{\xi'} \left\| \Theta \xi' - y^+ \right\|_2 + \alpha \left(\lambda \left\| \xi' \right\|_1 + \frac{(1-\lambda)}{2} \left\| \xi' \right\|_2^2 \right) \tag{12}$$

The λ controls the convex combination of L1 and L2. $\lambda = 1$ represents (12) is the LASSO algorithm, while $\lambda = 0$, (12) is the ridge regression algorithm. The weight α of the regularization is multiplied by the penalty term to determine the required sparse strength of the model. $\alpha = 0$ represents the least square method. The larger the α, the stronger the sparsity. The same thresholding design as STLS is added to the results of (12). By recursively calculating the elastic network results until the non-zero coefficients converge, STEN can get the convergent solution faster while ensuring sparsity.

4   Experimental Results and Analysis

The data are obtained by solving the unified functions of Splading, Nichols and Crocco-Busemann, and the regression experiment is carried out on the data. Splading function represents incompressible wall velocity formula. Nichols function represents the formula of compressible wall velocity. Crocco-Busemann function represents the compressible wall temperature formula. In engineering, these functions are very important for numerical simulation. By comparing the regression results of three algorithms introduced in

Sect. 3.1, it is found that STEN is more effective than LASSO and STLS in compressing coefficients. Comparing STEN algorithm with numerical calculation, STEN algorithm can ensure accuracy and spend less time.

Solving the wall function requires boundary conditions. The boundary conditions (including Mach number Ma, incoming temperature T_∞, wall temperature T_w, and wall pressure P_∞) are shown in Table 1.

Table 1. Numerical experiment parameter table

Case	Ma	T_∞	T_w	P_∞
1	3.0	108.5	271.0	21808.2
2	5.0	68.79	300.0	4079.89
3	8.324	71.4	294.3	14383.68
4	11.1	64.0	296.7	2541.5

The advantages and disadvantages of the STEN and LASSO algorithms are closely related to the regularization parameter λ. In the experiment, the results are compared by taking values from 0 to 1 at intervals of 0.1. STEN identifies the data of each example with good results. The performance of the regression model on the test set is shown in Fig. 1, 2 and 3. The coefficients obtained by the STEN algorithm are basically consistent with the coefficients of the theoretical function, and no extra function terms are regressed. The coefficient comparison is shown in Tables 2, 4, and 6. Regression coefficients of LASSO algorithm are quite different from the theoretical coefficients, and redundant function terms are regressed. The STLS algorithm has uncertainty in the performance of the wall function, it performs well in the Crocco-Busemann function, but poorly in the Splading and Nichols functions.

In this work, the algorithm models and numerical calculation procedures are written in Matlab programming language. The hardware environment is AMDRyzen 5 3550H, AMD Radeon Vega 8 Graphics, and the memory is 16 GB.

4.1 Splading Function

Merging similar terms for (1):

$$y^+ = -e^{-\kappa B} + \left(1 - \kappa e^{-\kappa B}\right)u^+ - \frac{\kappa^2 e^{-\kappa B}}{2}\left(u^+\right)^2 - \frac{\kappa^3 e^{-\kappa B}}{6}\left(u^+\right)^3 + e^{-\kappa B}e^{\kappa u^+} \quad (13)$$

Equation (13) contains the constant term, the first term, the second term, the third term, and the exponential term. Design candidate libraries $\Theta\left(u^+\right)$:

$$1, u^+, u^{+2}, u^{+3}, u^{+4}, u^{+5}, e^{\kappa u^+}, \sqrt{u^+}, \ln\left(y^+\right), e^{u^+}, u^{+1/3},$$
$$u^{+2/3}, u^{+3/2}, \sin\left(u^+\right), \cos\left(u^+\right), \ldots, \sin\left(4u^+\right), \cos\left(4u^+\right)$$

In order to illustrate the sparse performance of the algorithm, a variety of candidate terms are designed, including logarithmic term and trigonometric term besides its own function terms. 70% of the data are taken as the training set, 30% are used as the test set. After many experiments, when $\alpha = 0.1$, $\lambda = 0.1$, the regression model obtained by STEN is consistent with the theoretical model, and the remaining models have certain errors with the theoretical model. LASSO does not change much for different λ values, one of the better models is chosen to show the results. Regression results are shown in Table 2. The table only lists the terms whose coefficients are not zero in the regression results and the coefficients retain 4 decimal places.

Table 2. Splading regression results

Terms	Original coefficients	STEN	LASSO	STLS
1	−0.1049	−0.1049	−0.2322	−0.1049
u^+	0.9570	0.9570	1.2152	0.9570
u^{+2}	−0.0088	−0.0088	0	−0.0088
u^{+3}	−0.0012	−0.0012	0	−0.0012
$e^{\kappa u^+}$	0.1049	0.1049	0.0996	0.1049
$\sqrt{u^+}$	0	0	0	0.0005
$u^{+1/3}$	0	0	0	−0.0002
$u^{+2/3}$	0	0	0	−0.0004
$u^{+3/2}$	0	0	−0.1335	0

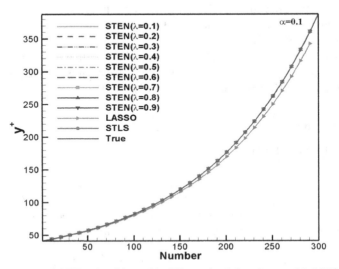

Fig. 1. Performance of STEN algorithm with different lambda values and LASSO and STLS algorithms on the test set of Splading function.

Table 2 shows that the regression model obtained by STEN is consistent with the theoretical model. Although the LASSO algorithm has strong compression ability, the regression coefficient error is large, and redundant function terms are generated. The STLS algorithm also generates redundant function terms as well as LASSO. The performance of each model on the test set is shown in Fig. 1.

Figure 1 shows that, except for the LASSO regression model, STEN and STLS are highly consistent with the test set. And the effect of STEN in the model is not sensitive to regular parameters. Compared with numerical calculation, STEN obtain a solution with higher accuracy, cost less time as the calculation time, shown in Table 3.

Table 3. Time consumption of splading

Number of data	Numeral calculations(s)	STEN(s)
1000	1.97E−03	1.14E−04
2000	2.29E−03	2.04E−04
3000	1.87E−03	1.68E−04
5000	2.32E−03	1.66E−04

4.2 Nichols and Nelson Function

By solving (2) with Case4 as the boundary condition to obtain data, and merging similar terms, we have:

$$y^+ = -e^{-\kappa B} + y^+_{white} + \left(1 - \kappa e^{-\kappa B}\right)u^+ - \frac{\kappa^2 e^{-\kappa B}}{2}\left(u^+\right)^2 - \frac{\kappa^3 e^{-\kappa B}}{6}\left(u^+\right)^3 \quad (14)$$

From the perspective of data, (2) is considered as a formula about two independent variables of y^+_{white} and u^+. The candidate function library Θ is designed as:

$$1, u^+, y^+_{white}, u^+ y^+_{white}, \ldots, u^{+^3}, y^{+^3}_{white}, e^{\kappa u^+}, \sqrt{u^+}, \ln\left(u^+\right), e^{u^+},$$
$$u^{+1/3}, u^{+2/3}, u^{+3/2}, \sin\left(u^+\right), \cos\left(u^+\right), \ldots, \sin\left(4u^+\right), \cos\left(4u^+\right)$$

We select 70% data as training set and 30% data as test set. For the Nichols equation, STEN performs well with $\alpha = 0.1$, $\lambda = 0.9$, and different λ in LASSO has little effect on the regression model. Thus, we select one of the situations randomly to show the results. The regression results are shown in Table 4.

Similar to the Splading regression results, the STEN algorithm performs well. The LASSO algorithm has high sparsity, but the generated function terms are not consistent with the objective function terms. STLS's ability to compress coefficients is weak, resulting in redundant function terms. The performance of each model in the test set is shown in Fig. 2.

The regression models obtained by STEN and STLS match the original data on the test set, while the LASSO model does not perform well. For time consumption, STEN has obvious advantages compared with numerical calculation.

Table 4. Nichols and Nelson regression results

Terms	Original coefficients	STEN	LASSO	STLS
1	−0.1049	−0.1049	−0.1745	−0.0199
u^+	0.9570	0.9570	0.7272	0.9083
y^+_{white}	1	1	0.4915	0.0885
u^{+^2}	−0.0088	−0.0088	0	−0.0064
u^{+^3}	−0.0012	−0.0012	0	−0.0011
$u^+ y^+_{white}$	0	0	0	0.0842
$y^+_{white}{}^2$	0	0	0.0148	0.0118
$e^{\kappa u^+}$	0	0	0	0.0104
$\sqrt{u^+}$	0	0	0	−0.0060
$u^{+^{1/3}}$	0	0	0	0.0019
$u^{+^{2/3}}$	0	0	0.3931	0.0062
$u^{+^{3/2}}$	0	0	0	0.0009

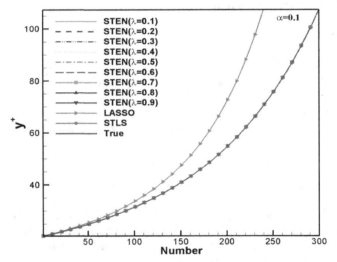

Fig. 2. Performance of STEN algorithm with different lambda values and LASSO and STLS algorithms on the test set of Nichols function.

There are few terms in y^+_{white} when designing candidate terms, because the value of y^+_{white} and y^+ are on the same order of magnitude. Through prior knowledge, we know them are weakly correlated. If we set a lot of terms about y^+_{white}, it will produce large error and affect the regression results. Therefore, we must grasp enough information about the regression data in order to set the candidates better. The candidates affect the recognition effect to a large extent. Experiments were carried out for each case in Table 1, but the more typical case4 was selected as an example to show the results.

Table 5. Time consumption of Nichols and Nelson

Number of data	Numeral calculations(s)	STEN(s)
1000	1.64E−03	1.82E−04
2000	1.76E−03	1.32E−04
3000	1.78E−03	1.29E−04
5000	1.86E−03	1.33E−04

4.3 Crocco-Busemann Function

The coupling coefficients Γ and β in (3) are fixed values in each calculation example, so they are not considered as variables from the perspective of data, but treated as coefficients. Design candidate library Θ:

$$1, u^+, u^{+2}, u^{+3}, u^{+4}, u^{+5}, e^{\kappa u^+}, \sqrt{u^+}, \ln(y^+), e^{u^+}, u^{+1/3},$$
$$u^{+2/3}, u^{+3/2}, \sin(u^+), \cos(u^+), \ldots, \sin(4u^+), \cos(4u^+)$$

STEN performs well with $\alpha = 0.0007, \lambda = 0.1$. The identification results with case2 are shown in Table 6.

Table 6. Crocco-Busemann regression results

Terms	Original coefficients	STEN	LASSO	STLS
1	1	1	0.9233	1
u^+	0.0106	0.0106	0	0.0106
u^{+2}	−0.0023	−0.0023	0	−0.0023

Table 7. Time consumption of Crocco-Busemann

Number of data	Numeral calculations(s)	STEN(s)
1000	4.95E−03	1.39E−04
2000	3.59E−03	1.32E−04
3000	4.39E−03	1.56E−04
5000	4.82E−03	1.46E−04

The three algorithms are excellent in compression coefficient, but the LASSO compression ability is too strong, resulting in underfitting. STEN and STLS both perform well. Different from the two examples, different λ values have obvious influence on STEN in Crocco-Busemann experiment. The error of LASSO algorithm is the biggest, followed by $\lambda = 0.5$ and $\lambda = 0.3$ in STEN. Similarly, the time cost of STEN is lower than that of numerical calculation, as shown in Table 7. The experiment was conducted for each Case in Table 1, The typical Case2 is selected as an example to show the results.

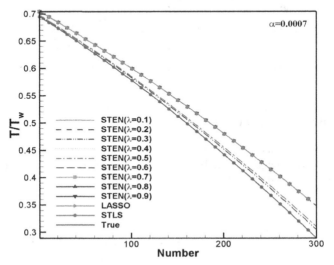

Fig. 3. Performance of STEN algorithm with different lambda values and LASSO and STLS algorithms on the test set of Crocco function.

5   Conclusion

An optimization algorithm based on elastic network called sequential threshold elastic network (STEN) is proposed. The STEN algorithm is applied to finding the functions from wall data. STEN can get the same function terms as the wall function, and its regression coefficients are basically the same. STEN algorithm is better than the LASSO and STLS algorithms both in the performance of the test set and the accuracy of the regression coefficients. And the time consumption compared to numerical calculation is less. STEN algorithm could be applied to various domain with ample data and the absence of governing equations, including neuroscience, climate science, and epidemiology. STEN algorithm could be applied to the grid data from experiments, then discovering the wall function which is more suitable for engineering application.

References

1. Jordan, M.I., Mitchell, T.M.: Machine learning: trends, perspectives, and prospects. Science **349**(6245), 255–260 (2015)

2. Vaddireddy, H., San, O.: Equation discovery using fast function extraction: a deterministic symbolic regression approach. Fluids **4**(2), 111 (2019)
3. Xu, H., Chang, H., Zhang, D.: DL-PDE: Deep-learning based data-driven discovery of partial differential equations from discrete and noisy data, arXiv preprint arXiv:1908.04463 (2016)
4. Raissi, M., Perdikaris, P., Karniadakis, G.E.: Physics-informed neural networks: a deep learning framework for solving forward and inverse problems involving nonlinear partial differential equations. J. Comput. Phys. **378**, 686–707 (2019)
5. Raissi, M.: Deep hidden physics models: deep learning of nonlinear partial differential equations. J. Mach. Learn. Res. **19**(1), 932–955 (2018)
6. Long, Z., Yiping, L., Dong, B.: PDE-Net 2.0: learning PDEs from data with a numeric-symbolic hybrid deep network. J. Comput. Phys. **399**, 108925 (2019)
7. Long, Z., Lu, Y., Ma, X., Dong, B.: PDE-Net: learning PDEs from data. In: International Conference on Machine Learning, pp. 3208–3216 (2018)
8. Haber, E., Ruthotto, L.: Stable architectures for deep neural networks. Inverse Prob. **34**(1), 014004 (2017)
9. Brunton, S.L., Noack, B.R., Koumoutsakos, P.: Machine learning for fluid mechanics. Ann. Rev. Fluid Mech. **52**(1), 477–508 (2020)
10. Brunton, S.L., Proctor, J.L., Kutz, J.N.: Discovering governing equations from data by sparse identification of nonlinear dynamical systems. Proc. Natl. Acad. Sci. **113**(15), 3932–3937 (2016)
11. Rudy, S.H.: Data-driven discovery of partial differential equations. Sci. Adv. **3**(4), e1602614 (2017)
12. Rudy, S.: Data-driven identification of parametric partial differential equations. SIAM J. Appl. Dyn. Syst. **18**(2), 643–660 (2019)
13. Schaeffer, H.: Learning partial differential equations via data discovery and sparse optimization. Proc. Royal Soc. A Math. Phys. Eng. Sci. **473**(2197), 20160446 (2017)
14. Zhang, J., Ma, W.: Data-driven discovery of governing equations for fluid dynamics based on molecular simulation. J. Fluid Mech. **892** (2020)
15. Mangan, N.M.: Inferring biological networks by sparse identification of nonlinear dynamics. IEEE Trans. Mol. Biol. Multi-Scale Commun. **2**(1), 52–63 (2016)
16. Schaeffer, H., McCalla, S.G.: Sparse model selection via integral terms. Phys. Rev. E **96**(2), 023302 (2017)
17. Boninsegna, L., Nüske, F., Clementi, C.: Sparse learning of stochastic dynamical equations. J. Chem. Phys. **148**(24), 241723 (2018)
18. Schaeffer, H., Tran, G., Ward, R.: Extracting sparse high-dimensional dynamics from limited data. SIAM J. Appl. Math. **78**(6), 3279–3295 (2018)
19. Berg, J., Nyström, K.: Data-driven discovery of PDEs in complex datasets. J. Comput. Phys. **384**, 239–252 (2019)
20. Chang, H., Zhang, D.: Identification of physical processes via combined data-driven and data-assimilation methods. J. Comput. Phys. **393**, 337–350 (2019)
21. Loiseau, J.-C., Brunton, S.L.: Constrained sparse Galerkin regression. arXiv preprint arXiv: 1611.03271 (2016)
22. Li, X.: Sparse learning of partial differential equations with structured dictionary matrix. Chaos Interdisc. J. Nonlinear Sci. **29**(4), 043130 (2019)
23. Spalding, D.B.: A single formula for the law of the wall. J. Appl. Mech. **28**(3), 455–458 (1961)
24. Nichols, R.H., Nelson, C.C.: Wall function boundary conditions including heat transfer and compressibility. AIAA J. **42**(6), 1107–1114 (2004)
25. White, F.M., Christoph, G.H.: A simple new analysis of compressible turbulent two-dimensional skin friction under arbitrary conditions, Rhode Island Univ Kingston Dept of Mechanical Engineering and Applied Mechanics (1971)

26. Tibshirani, R.: Regression shrinkage and selection via the lasso. J. Royal Stat. Soc. Ser. B (Methodological) **58**(1), 267–288 (1996)
27. Hans, C.: Elastic net regression modeling with the orthant normal prior. J. Am. Stat. Assoc. **106**(496), 1383–1393 (2011)
28. Marx, V.: Biology: the big challenges of big data. Nature **498**(7453), 255–260 (2013)

Rotation-DPeak: Improving Density Peaks Selection for Imbalanced Data

Xiaoliang Hu[1,2,3], Ming Yan[1,2,3], Yewang Chen[1,2,3(✉)], Lijie Yang[1], and Jixiang Du[1,3]

[1] The College of Computer Science and Technology, Huaqiao University, Xiamen, China
ywchen@hqu.edu.cn
[2] Provincial Key Laboratory for Computer Information Processing Technology, Soochow University, Soochow, China
[3] Fujian Key Laboratory of Big Data Intelligence and Security, Huaqiao University, Xiamen, China

Abstract. Density Peak (DPeak) is an effective clustering algorithm. It maps arbitrary dimensional data onto a 2-dimensional space, which yields cluster centers and outliers automatically distribute on upper right and upper left corner, respectively. However, DPeak is not suitable for imbalanced data set with large difference in density, where sparse clusters are usually not identified. Hence, an improved DPeak, namely Rotation-DPeak, is proposed to overcome this drawback according to an simple idea: the higher density of a point p, the larger δ it should have such that p can be picked as a density peak, where δ is the distance from p to its nearest neighbor with higher density. Then, we use a quadratic curve to select points with the largest decision gap as density peaks, instead of choosing points with the largest γ, where $\gamma = \rho \times \delta$. Experiments shows that the proposed algorithm obtains better performance on imbalanced data set, which proves that it is promising.

Keywords: Density peak · Decision curve · Decision gap

1 Introduction

Clustering, as known as unsupervised classification, automatically divides the data set into different groups or more subsets according to measured or intrinsic characteristic or similarity. It is a crucial technique for many fields ranging from Bioinformatics to image processing and social network, and various clustering algorithms have been proposed in the past 50 years, and mainly be classified into several categories, including partition-based [1], hierarchical [2], grid-based [3], density-based [4–7], graph-based [8] and some others [9–12].

DPeak [4] is a density-based clustering algorithm, and can recognize clusters of arbitrary shapes. It has a basis in the assumptions that cluster centers are

https://github.com/XFastDataLab/Rotation-DPeak.

© Springer Nature Singapore Pte Ltd. 2021
H. Mei et al. (Eds.): BigData 2020, CCIS 1320, pp. 45–58, 2021.
https://doi.org/10.1007/978-981-16-0705-9_4

surrounded by neighbors with lower local density and that they are at a relatively large distance from any points with a higher local density.

Let P be a data set, n be the cardinality of P, $d_{i,j}$ be the distance from point i to j. For each data point i, DPeak computes two quantities: its local density ρ_i and the distance δ_i from its nearest neighbor point of higher local density, as below:

$$\rho_i = \sum_j \chi(d_{ij} - d_c) \tag{1}$$

where $\chi(x) = 1$ if $x < 0$ else $\chi(x) = 0$, and d_c is a cutoff distance (threshold) which is predefined by user.

δ_i is the distance between point i and its nearest neighbor with higher density:

$$\delta_i = \min_{j:\rho_j > \rho_i} d_{ij} \tag{2}$$

For the point with the highest density, its δ is denoted as $\delta_i = max_j(d_{ij})$.

DPeak has attracted many attentions and been applied in many fields, such as image processing [13], community detection [14,15], extracting muti-document abstracts [16] and noise removal [17], because of its simplicity and intuitiveness. However, it is not suitable for imbalanced data, e.g., gene sequence reorganization [18], which has large different density distributions in different regions, due to that DPeak prefers selecting peaks from regions with high density instead of sparse regions, yielding sparse region is usually mis-classified into other cluster.

In this paper, a novel algorithm, namely Rotation-DPeak, is proposed, which outperforms DPeak and other variants of DPeak on nornal and imbalanced data. The main contributions of this paper are: (1) A novel idea "the higher density of a point p, the larger δ it should have such that p can be picked as a density peak" is proposed to overcome the disadvantage of DPeak on imbalance data; (2) A new strategy of choosing density peaks is using a quadratic curve to select points with the largest decision gap.

The rest of this paper is organized as follows: Sect. 2 lists related work; Sect. 3 presents the drawbacks of the method for selecting density peaks in original DPeak algorithm; In Sect. 4, the proposed algorithm is introduced; In Sect. 5, we conduct experiments on various data sets, including normal and imbalanced data, and analysis the results; Sect. 6 draws the conclusion.

2   Related Work

By decision graph, DPeak [4] provides two ways to select the density peaks. One is to use a rectangle. The other is to choose first K number of points with the largest γ, where $\gamma = \rho \times \delta$, which is in fact choosing points upon a hyperbolic curve to be peaks. However, in some cases, even if the parameter d_c is well chosen, it is still difficult to pick out the optimal density peaks from the decision graph, especially for imbalanced data.

Since the birth of DPeak, extensive works over the past six years have produced a rich collection of algorithms to discuss, accelerate and apply it in various applications [19] [20], etc.,. One of the most popular topics is to promote the performance of finding better the density peaks.

Based on an underlying idea that each cluster has a shrunken density core region that roughly retains the shape of the cluster, Chen et al. [21] proposed DCore to find a loose and distributed density core for each cluster, instead of selecting a peak. ADPC-kNN [22] uses k-nearest neighbors to compute the global parameter d_c and the local density ρ_i of each point, applies a new approach to select initial cluster centers automatically, and finally aggregate clusters if they are density reachable. Motivated by the divide-and-conquer strategy and the density-reachable concept, 3DC algorithm [23] was designed find the correct number of clusters in a recursive way. Wang et al. [24] proposed a clustering algorithm, namely ADPclust, with adaptive density peak detection, where the local density is estimated through the nonparametric multivariate kernel estimation, and developed an automatic cluster centroid selection method through maximizing an average silhouette index. DPC-GEV [25] uses the judgment index which equals the lower value within density and distance (after normalization) to select the clustering centers. The judgment index approximately follows the generalized extreme value (GEV) distribution, and each clustering center's judgment index is much higher.

3   Deficiency of DPeak for Choosing Density Peak

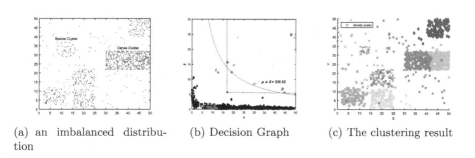

(a) an imbalanced distribution   (b) Decision Graph   (c) The clustering result

Fig. 1. An example: (a) plots an imbalanced data set with a dense region and a sparse region. (b) shows that all points upon the red hyperbolic curve (with equation $\rho \times \delta = 235.62$) or within the rectangle are selected as density peaks; all points in the sparse region are marked red, and t is the point with the largest γ among all them. (c) shows the clustering result of the DPeak algorithm. The sparse region is merged into another cluster, and the dense region is divided into two classifications. (Color figure online)

As mention above, DPeak provides two methods to select cluster centers. One is to choose the first K number of points with the largest γ to be density peaks, where K is a user specified cluster number and $\gamma = \rho \times \delta$. In fact, this way works in a way that picks density peaks according to a hyperbolic curve. Another is to pick density peaks within a rectangle.

For example, as Fig. 1 shows, (a) shows the distribution of an imbalanced data, i.e., there are some dense and sparse clusters, (b) plots the decision graph, where we can see that there are 6 points, which distribute upon the red hyperbolic curve and also within the rectangle as well, are picked as density peaks. The red curve is in fact a hyperbolic curve with equation $\rho \times \delta = 235.62$, where 235.62 is the lowest γ value of the 6 density peaks.

However, this method has a deficiency in some cases, as follows. Suppose P is an imbalanced data set, $p \in P$ is a point in a dense region and $t \in P$ is another point in a very sparse region. Then, usually p may have a higher density than that of t, e.g., 10 or more times larger than that of t. Meanwhile, in the case of δ of two points are similar, which may yields that the γ of p is larger than that of t. Thus, p has a higher possibility than t to be selected as a density peak. Similarly, if the unit of δ is far larger than of ρ, some noise points may be mis-classified as density peaks.

For example, as shown in Fig. 1(a), the data set should be classified into 6 clusters naturally from naked eye. But in Fig. 1(b) and (c), it is observed that the result is not as good as we expect, because t should be a density peak for a sparse cluster, while p should not. In (b), we can see the density of t is far less than p, but the difference of δ between them is not so obvious, yielding that the γ of t is less than that of p. In the case that the cluster number $K = 6$, t fails to be selected as a density peak, and the whole sparse region is merged into another cluster, while p is wrongly picked out which results in dividing the dense region into two clusters.

4   The Proposed Algorithm

Since there is a drawback in DPeak clustering algorithm to select density peaks for imbalanced data set. Hence, in this section, we will propose a novel method to overcome it based on a new idea as below.

4.1   Our Idea

As mentioned in above, the original DPeak algorithm picks density peaks by a hyperbolic curve or a rectangle, which may yields unsatisfied performance for imbalanced data set, due to that the cluster center of the sparse region usually has relatively low γ.

Therefore, we propose a simple idea: the higher density of a point p, the larger δ it should have such that it can be picked as a density peak.

As we can see from Fig. 1 the hyperbolic curve is unable to pick density peaks from sparse regions, due to that it only takes γ into consideration. Hence, quadratic curve is optional, as shown in Fig. 2(a), it is observed that the quadratic curve tails up in the right bottom corner. It implies a point with high density should have larger δ such that it can be picked as a density peak. While for a point with relatively low density, it still has a great opportunity to be a density peak if it has a relatively large δ. For example, we obtain better

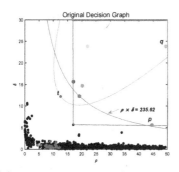 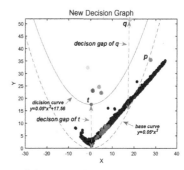

(a) Quadratic curve VS Hyperbolic curve and Rectangle

(b) Rotated Decision Graph

Fig. 2. (a) makes a comparison for quadratic curve to hyperbolic curve and rectangle, the former selects density peaks including t without p, while the latter two methods have opposite choice. (b) shows the new decision graph after rotation. The red dash curve is base curve, and blue quadratic parabola curve is the so called decision curve with equation $y = 0.05 * x^2 + 17.56$, which is determined by selecting first 6 largest decision gap. q has the largest decision gap, and the decision gap of p is obviously less than that of t. All points upon the decision curve are density peaks. (Color figure online)

result in this figure, for that p is filtered and t is picked out as a density peak according to this new standard.

4.2   Data Prepossession and Rotation

Because the unit of ρ and δ are quite different, in order to eliminate the influence of unit, we normalize ρ and δ, by using Min-Max normalization for two quantities as below.

$$\rho_i^{new} = \frac{\rho_i - \rho_{min}}{\rho_{max} - \rho_{min}} \tag{3}$$

$$\delta_i^{new} = \frac{\delta_i - \delta min}{\delta_{max} - \delta_{min}} \tag{4}$$

As mentioned above, quadratic curve is suitable for solving the problem of hyperbolic curve. But it is not convenient for us to use a quadratic equation directly under the ρ-δ coordinates. Hence, we would like to make counter clockwise rotation on the decision graph as below.

Definition 1 (Rotation Matrix). *The **Rotation Matrix** is a matrix, which rotates the column vector counter clockwise by θ, i.e.,*

$$T(\theta) = \begin{bmatrix} cos\theta & -sin\theta \\ sin\theta & cos\theta \end{bmatrix} \tag{5}$$

Definition 2 (Representative Vector). *The* **Representative Vector** *of point i is a vector containing* ρ *and* δ: $V_i = \begin{bmatrix} \rho_i \\ \delta_i \end{bmatrix}$

Let $NV_i = \begin{bmatrix} x_i \\ y_i \end{bmatrix}$ be a new vector obtained by using $T(\theta)$ to rotate, i.e.,:

$$NV_i = \begin{bmatrix} x_i \\ y_i \end{bmatrix} = T(\theta) \times V_i \tag{6}$$

As shown in Fig. 2(b), in this figure we rotate the whole decision graph counter clockwise by $T(\frac{\pi}{4})$. Each point has a one-to-one correspondence to a point in Fig. 2(a). Now, the quadratic equation can be easily described as:

$$y = a \times x^2 + c \tag{7}$$

where a and c are two parameters.

Definition 3 (Base Curve). *The parabola of* $y = a \times x^2$ *is called Base Curve w.r.t a.*

As the red dash parabola shows in Fig. 2(b), it is a base curve w.r.t 0.05.

Definition 4 (Decision Curve) *The parabola of quadratic Eq. (7) is called Decision Curve w.r.t a and c.*

Definition 5 (Decision Gap). *We call* $\eta_i = y_i - a \times x_i^2$ *the decision gap of point i w.r.t a.*

As the two dash segment line present in Fig. 2(b), one is the decision gap of t w.r.t 0.05, and the other is the decision gap of q w.r.t 0.05.

4.3   Strategy for Selecting Density Peaks

So far, we can also provide two ways of selecting density peaks as below:

Strategy (1): The rule for selecting point i as a peak is that i should satisfy:

$$y_i \geq a \times x_i^2 + c \tag{8}$$

where a and c are two user specified parameters. In this case, the number of clusters found is automatically determined.

Definition 6 (Decision Equation). $y \geq a \times x^2 + c$ *is called Decision Equation of decision curve w.r.t a and c.*

Strategy (2): The standard is to choose the first K number of points with the largest gap i.e., η. Let t be the point with the Kth largest decision gap, then the equation of the decision curve which passes through t is:

$$y = a \times x^2 + y_t \tag{9}$$

Hence, its **Decision Equation** is

$$y \geq a \times x^2 + y_t \tag{10}$$

For example, let $a = 0.05$ and $K = 6$, in Fig. 2(b) $\eta_t = y_t - 0 = 17.56$ is the 6th largest gap, then, the **Decision Equation** is $y \geq 0.05 \times x^2 + 17.56$. Hence, in this case $p = [27.55, 35.48]^T$ fails to be selected, due to that it falls under the decision curve, i.e., $y_p = 35.48 < 0.05 \times (27.55)^2 + 17.56$.

Algorithm 1. Rotation-DPeak

Input: Local Density Set; Distance Matrix D; data Set P_1;d_c; a, K; $\theta = \frac{\pi}{4}$
Output: cluster id of each point CL;
  1: Initializing: Density Peak Set $DPS = \phi$;
  2: Compute ρ and δ for each point by original DPeak with d_c
  3: DPS = select density peaks by Algorithm 2 with a and K
  4: $Ncluster = 1$
  5: **for** each point $j \in DPS$ **do**
  6:     $CL(j) = Ncluster$
  7:     $Ncluster = Ncluster + 1$
  8: **end for**
  9: assign label for other points in the same way as original DPeak

---

4.4 Algorithm

Hence, we propose a novel DPeak algorithm(Rotation-DPeak). The main steps are: it computes ρ and δ by original DPeak, and selects density peaks by Algorithm 2, then assigns cluster label for all non-peak points in the same way as DPeak.

The key procedure is Algorithm 2. As mentioned previously, it rotates the decision graph counter clockwise, and picks out density peaks by selecting the first K number of points with the largest decision gap, (here, we only present the second strategy of selecting density peaks mentioned above).

Complexity Analysis: In Algorithm 2, the complexity from line 6 to line 9 is $O(n)$, line 10 is about $O(n \, logn)$. Hence, Algorithm 2 runs in $O(n \, logn)$ expected time. In Algorithm 1, due to that the complexity of original DPeak is $O(n^2)$, it runs in $O(n^2)$ expected time.

5 Experiments

In order to evaluate the performance of the proposed algorithm, we conduct experiments on some normal data sets from UCI [1], and some synthetic imbalanced data sets. Both DPeak and the proposed algorithm are written in Matlab R2012a.

---

[1] https://archive.ics.uci.edu/ml/index.php.

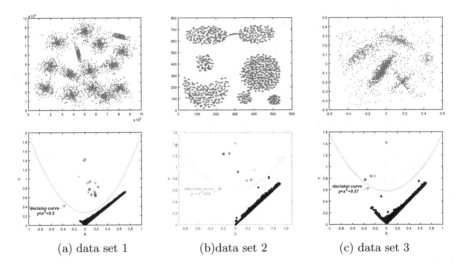

(a) data set 1   (b)data set 2   (c) data set 3

Fig. 3. Three examples of new decision graph on three data sets: In data set 1 and 2, we use *strategy (1)* to select peaks, specify $a = 1$, $c = 0.5$ and $a = 1$, $c = 0.6$, i.e., decision equations are $y \geq x^2 + 0.5$ and $y \geq x^2 + 0.6$, respectively; In (c), *strategy (2)* is applied with $a = 1, K = 5$, yielding the decision equation $y \geq x^2 + 0.57$.

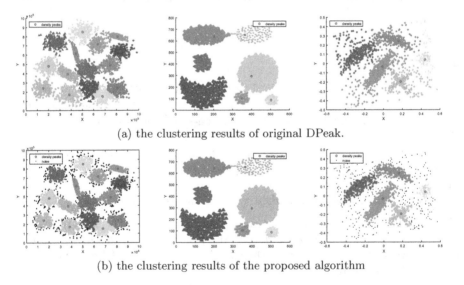

(a) the clustering results of original DPeak.

(b) the clustering results of the proposed algorithm

Fig. 4. The clustering results of DPeak and Rotation-DPeak on three normal data sets.

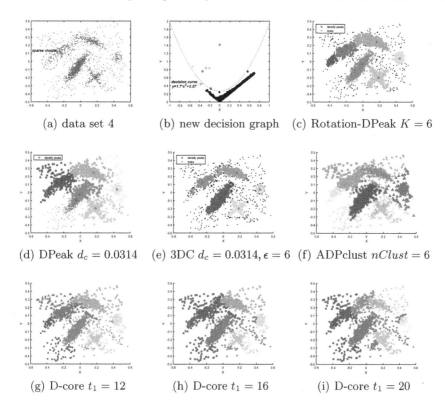

(a) data set 4   (b) new decision graph   (c) Rotation-DPeak $K = 6$

(d) DPeak $d_c = 0.0314$ (e) 3DC $d_c = 0.0314, \epsilon = 6$ (f) ADPclust $nClust = 6$

(g) D-core $t_1 = 12$ (h) D-core $t_1 = 16$ (i) D-core $t_1 = 20$

Fig. 5. An example on imbalanced data: (a) The data distribution of data set 4; (b)The new decision graph with $d_c = 0.0314, \theta = \frac{\pi}{4}, K = 6$, yielding the decision equation $y \geq 1.7 * x^2 + 0.27$; (c)The clustering result of the proposed algorithm; (d)–(i) are the clustering result of DPeak, 3DC, ADPclust and D-core, respectively.

5.1 Experiments on Normal Data Sets

In this subsection, we conduct some experiments on 2-dimensional synthetic data sets to demonstrate the generality of our approach, and make comparisons with original DPeak [4]. As shown in Fig. 3, three data sets were designed to consider different degrees of cluster density, the various cluster patterns and different cluster sizes in order to evaluate the proposed algorithm and DPeak.

The parameter d_c is set to be 2% quantile of all distances among points in the data set. i.e., $d_c = distVec[\lceil 2\% \times \frac{n^2-n}{2} \rceil]$ where $distVec$ is a sorted distance vector for all distances, n is the total number of points, and $\lceil \cdot \rceil$ is a function that trims a real to be an integer. As far as θ is considered, in most cases we believe that ρ and δ are equally important, hence, we adopt $\theta = \frac{\pi}{4}$ as default value.

The first column of Fig. 3 shows the distribution of three data sets, and the second column demonstrates three new decision graphs on them, respectively. In data set 1 and 2, we use strategy (1), while in data set 3 strategy (2) is applied. The clustering results of DPeak and the proposed method are plotted

---

Algorithm 2. Peak-Selection

Input: ρ Set; δ set; data Set P_1; a, K, θ
Output: Density Peak Set DPS;
1: Initializing:$DPS = \phi$;
2: $T(\theta) = \begin{bmatrix} cos\theta & -sin\theta \\ sin\theta & cos\theta \end{bmatrix}$
3: ρ^{new} = normalize ρ by Equation (3)
4: δ^{new} = normalize δ by Equation (4)
5: $V_i = \begin{bmatrix} \rho_i^{new} \\ \delta_i^{new} \end{bmatrix}$
6: **for** each point $i \in P_1$ **do**
7:     $NV_i = T(\theta) \times V_i$
8:     $\eta_i = NV_i[2] - a \times NV_i[1]^2$ //decision gap
9: **end for**
10: sort η
11: DPS= the first K number of points with the largest η

---

in Fig. 4(a) and (b). We can see that on these normal data sets, both algorithms can correctly discover the clusters and detect outliers. In Fig. 4(a) to (f), it is observed that both DPeak and Rotation-DPeak obtain correct results.

5.2   Experiments on Imbalanced Data Sets

In this subsection, we conduct some experiments on imbalanced data sets, and make comparisons with DPeak [4], 3DC [23], ADPClust [24] and D-core [21]. 3DC has two parameters, one is $Parameter.percent$ which is used to compute cutoff distance d_c in DPeak, the other is $Parameter.d_c$ which is a distance threshold for judging whether one point is density reachable from another, i.e., it plays the same role as ϵ in DBSCAN. In order to avoid ambiguity, we use character ϵ instead of $Parameter.d_c$ below. In ADPclust, there is a parameter $nClut$ which specifies the cluster number, i.e. it equals the parameter K of the proposed method. D-core has three parameters, which are r_1, ClNum, t_1. r_1 is a distance metric that is used to calculate density, playing the same role as $Parameter.d_c$ in DPeak. ClNum is the cluster number which is equal to K in the proposed method. t_1 is a density threshold.

Data set 4 is a data set revised from data set 3 by removing some points, which makes a region become sparse, as the red dash circle shows in Fig. 5(a). It is reasonable to classify this data set into 6 clusters. Hence, we set $a = 1.7$, $K = 6$ and $d_c = 0.0314$. Then, we obtain the final clustering result in Fig. 5(c). While the results of original DPeak, 3DC, ADPclust and D-core are not as good as the proposed algorithm. DPeak is easy to mis-classify the dense region into two parts no mater d_c, as Fig. 5(d) show; 3DC only yields 5 clusters as presented in Fig. 5(e); ADPclust has many errors for classifying border and noise points, as demonstrated in Fig. 5(f), and D-core mis-classify some noises into clusters, as Fig. 5(g) to Fig. 5(i) shown.

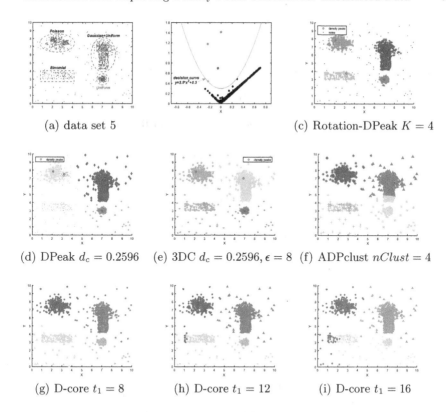

(a) data set 5   (c) Rotation-DPeak $K = 4$

(d) DPeak $d_c = 0.2596$ (e) 3DC $d_c = 0.2596, \epsilon = 8$ (f) ADPclust $nClust = 4$

(g) D-core $t_1 = 8$ (h) D-core $t_1 = 12$ (i) D-core $t_1 = 16$

Fig. 6. Example on imbalanced data set 5: (a) the point distribution; (b) the new decision graph with $d_c = 0.2596, \theta = \frac{\pi}{4}, K = 4$, yielding the decision equation $y \geq 2.5 * x^2 + 0.3$; (c) the clustering result of the proposed algorithm; (d)–(i) are the clustering result of DPeak, 3DC, ADPclust and D-core, respectively.

Data set 5 and data set 6 are both synthetic data sets, which are made of some clusters with quite different density distributions (such as Gaussian, Uniform and Poisson distribution, etc.). As Fig. 6 and Fig. 7 show, the proposed algorithm still outperforms DPeak, 3DC, ADPclust or D-core, for these competitors do not work well on such imbalanced data.

Comprehensive Analysis: Due to limitation of pages, we only give one result for each algorithm above. In fact, no mater how the parameters vary, DPeak and its current variants are not suitable for identifying sparse clusters from imbalanced data. From these experiments, we can see that Rotation-DPeak outperforms DPeak on imbalanced data set, due to that its basic strategy for selecting peaks is not too biased towards γ value.

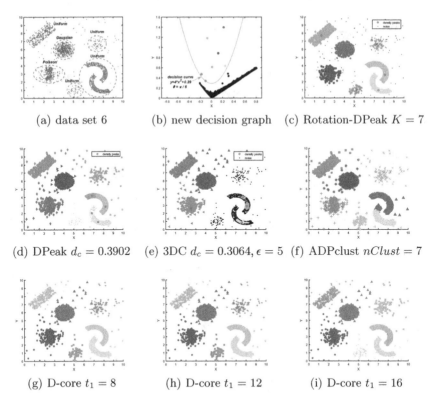

(a) data set 6   (b) new decision graph   (c) Rotation-DPeak $K = 7$

(d) DPeak $d_c = 0.3902$ (e) 3DC $d_c = 0.3064, \epsilon = 5$ (f) ADPclust $nClust = 7$

(g) D-core $t_1 = 8$ (h) D-core $t_1 = 12$ (i) D-core $t_1 = 16$

Fig. 7. Example on imbalanced data set 6: (a) the data distribution; (b) the new decision graph with $d_c = 0.4672, \theta = \frac{\pi}{6}, K = 7$, yielding the decision equation $y \geq 4 * x^2 + 0.29$; (c) the clustering result of the proposed algorithm; (d)–(i) are the clustering result of DPeak, 3DC, ADPclust and D-core, respectively.

6   Conclusion

In this paper, we propose Rotation-DPeak clustering algorithm based on a simple idea:'the higher density of a point p, the larger δ it should have such that p can be picked as a density peak'. Rotation-DPeak rotates the decision graph, and uses a quadratic parabola decision curve to select points with the largest decision gap as density peaks, rather than choosing points with the largest γ in original DPeak. The experimental results on the normal and imbalanced data demonstrate that the Rotation-DPeak algorithm outperforms four other algorithms referenced in this paper.

Due to that the parameter K of Rotation-DPeak is specified by users, in future work, we will attempt to invent a new strategy to automatically determine K. Meanwhile, we find that some data points with high density in dense region of an imbalanced data may be classified as Outliers. Therefore, we will do some researches on detecting outliers.

Acknowledgment. We acknowledge financial support from the National Natural Science Foundation of China (No. 61673186, 61972010, 61975124).

References

1. Likas, A., Vlassis, N., Verbeek, J.J.: The global k-means clustering algorithm. Pattern Recognit. **36**(2), 451–461 (2003)
2. Zhong, C., Miao, D., FrNti, P.: Minimum spanning tree based split-and-merge: a hierarchical clustering method. Inf. Ences **181**(16), 3397–3410 (2011)
3. Wang, W., Yang, J., Muntz, R.: Sting: a statistical information grid approach to spatial data mining. In: Proceedings of 23rd International Conference Very Large Data Bases, VLDB 1997, Athens, Greece, pp. 186–195 (1997)
4. Rodriguez, A., Laio, A.: Clustering by fast search and find of density peaks. Science **344**(6191), 1492–1496 (2014)
5. Chen, Y., Tang, S., Bouguila, N., Wang, C., Du, J., Li, H.: A fast clustering algorithm based on pruning unnecessary distance computations in DBSCAN for high-dimensional data. Pattern Recognit. **83**, 375–387 (2018)
6. Chen, Y., et al.: KNN-block DBSCAN: fast clustering for large-scale data. IEEE Trans. Syst. Man Cybern. Syst. 1–15 (2019)
7. Chen, Y., Zhou, L., Bouguila, N., Wang, C., Chen, Y., Du, J.: Block-DBSCAN: fast clustering for large scale data. Pattern Recognit. **109**, 107624 (2021)
8. Kang, Z., Wen, L., Chen, W., Xu, Z.: Low-rank kernel learning for graph-based clustering. Knowl. Based Syst. **163**, 510–517 (2019)
9. Kang, Z., et al.: Partition level multiview subspace clustering. Neural Netw. **122**, 279–288 (2020)
10. Xing, Y., Yu, G., Domeniconi, C., Wang, J., Zhang, Z., Guo, M.: Multi-view multi-instance multi-label learning based on collaborative matrix factorization, pp. 5508–5515 (2019)
11. Huang, D., Wang, C.D., Wu, J., Lai, J.H., Kwoh, C.K.: Ultra-scalable spectral clustering and ensemble clustering. IEEE Trans. Knowl. Data Eng. **32**(6), 1212–1226 (2019)
12. Zhang, Z., et al.: Flexible auto-weighted local-coordinate concept factorization: a robust framework for unsupervised clustering. IEEE Trans. Knowl. Data Eng. 1 (2019)
13. Shi, Y., Chen, Z., Qi, Z., Meng, F., Cui, L.: A novel clustering-based image segmentation via density peaks algorithm with mid-level feature. Neural Comput. Appl. **28**(1), 29–39 (2016). https://doi.org/10.1007/s00521-016-2300-1
14. Bai, X., Yang, P., Shi, X.: An overlapping community detection algorithm based on density peaks. Neurocomputing **226**(22), 7–15 (2017)
15. Liu, D., Su, Y., Li, X., Niu, Z.: A novel community detection method based on cluster density peaks. In: National CCF Conference on Natural Language Processing & Chinese Computing, vol. PP, pp. 515–525 (2017)
16. Wang, B., Zhang, J., Liu, Y.: Density peaks clustering based integrate framework for multi-document summarization. CAAI Trans. Intell. Technol. **2**(1), 26–30 (2017)
17. Li, C., Ding, G., Wang, D., Yan, L., Wang, S.: Clustering by fast search and find of density peaks with data field. Chin. J. Electron. **25**(3), 397–402 (2016)

18. Mehmood, R., El-Ashram, S., Bie, R., Sun, Y.: Effective cancer subtyping by employing density peaks clustering by using gene expression microarray. Pers. Ubiquit. Comput. **22**(3), 615–619 (2018). https://doi.org/10.1007/s00779-018-1112-y

19. Cheng, D., Zhu, Q., Huang, J., Wu, Q., Lijun, Y.: Clustering with local density peaks-based minimum spanning tree. IEEE Trans. Knowl. Data Eng. PP(99), 1 (2019). https://doi.org/10.1109/TKDE.2019.2930056

20. Chen, Y., et al.: Fast density peak clustering for large scale data based on KNN. Knowl. Based Syst. **187**, 104824 (2020)

21. Chen, Y., et al.: Decentralized clustering by finding loose and distributed density cores. Inf. Sci. **433–434**, 649–660 (2018)

22. Yaohui, L., Zhengming, M., Fang, Y.: Adaptive density peak clustering based on k-nearest neighbors with aggregating strategy. Knowl. Based Syst. **133**, 208–220 (2017)

23. Liang, Z., Chen, P.: Delta-density based clustering with a divide-and-conquer strategy: 3DC clustering. Pattern Recognit. Lett. **73**, 52–59 (2016)

24. Wang, X.F., Xu, Y.: Fast clustering using adaptive density peak detection. Stat. Methods Med. Res. **26**(6), 2800–2811 (2017)

25. Ding, J., He, X., Yuan, J., Jiang, B.: Automatic clustering based on density peak detection using generalized extreme value distribution. In: Soft Computing. A Fusion of Foundations Methodologies & Applications, pp. 515–525 (2018)

Introducing MDPSD, a Multimodal Dataset for Psychological Stress Detection

Wei Chen, Shixin Zheng, and Xiao Sun[✉]

Hefei University of Technology, Hefei 230009, China
`shixinz@mail.hfut.edu.cn`, `sunx@hfut.edu.cn`

Abstract. As we all know, long-term stress can have a serious impact on human health, which requires continuous and automatic stress monitoring systems. However, there is a lack of commonly used standard data sets for psychological stress detection in affective computing research. Therefore, we present a multimodal dataset for the detection of human stress (MDPSD). A setup was arranged for the synchronized recording of facial videos, photoplethysmography (PPG), and electrodermal activity (EDA) data. 120 participants of different genders and ages were recruited from universities to participate in the experiment. The data collection experiment was divided into eight sessions, including four different kinds of psychological stress stimuli: the classic Stroop Color-Word Test, the Rotation Letter Test, the Stroop Number-Size Test, and the Kraepelin Test. Participants completed the test of each session as required, and then fed back to us the self-assessment stress of each session as our data label. To demonstrate the dataset's utility, we present an analysis of the correlations between participants' self-assessments and their physiological responses. Stress is detected using well-known physiological signal features and standard machine learning methods to create a baseline on the dataset. In addition, the accuracy of binary stress recognition achieved 82.60%, and that of three-level stress recognition was 61.04%.

Keywords: Stress detection · Physiological signals · Facial videos · Affective computing · Data process

1 Introduction

In modern life, people will have psychological stress, from the study, life, and work. Chronic stress can lead to mental illness, such as generalized anxiety disorder and depression [10]. Stress has a multidimensional effect that affects every level of human activity, and it is not always possible to estimate the primary effect or intensity of the effect. Physiological stress response refers to physical changes caused by environmental events or conditions, called stressors.

Supported by General Program of National Natural Science Foundation of China (61976078).

H. Mei et al. (Eds.): BigData 2020, CCIS 1320, pp. 59–82, 2021.
https://doi.org/10.1007/978-981-16-0705-9_5

This response involves the following physiological processes: processing potential stressors and organizing adaptive responses; mobilize the musculoskeletal system to prepare and execute motor movements. This work focused on physiological changes of photoplethysmography (PPG), electrodermal activity (EDA), and facial information in the body's response to stressors.

Psychometric responses may depend on the mood and mental state of the participants on the day of participating study [1]. In this study, physiological measures were chosen as the main method because participants could not consciously manipulate their autonomic nervous system (ANS) activity [6]. In addition, physiological measurements provide a non-invasive method for determining the stress levels and responses of participants interacting with tests. Even if the psychometrics did not provide a significant correlation between stress and physiological responses, the physiological measurements provided an indication of physiological changes associated with stress-related changes. Common stressors include many categories. We carried out the research based on the mental or tasks related stressor. The physiological response of the body should be the same under the stimulation of different stressors. In this way, we can make the subjects generate a certain amount of stress through the artificial stressor stimulation, and collect the physiological data of the subjects under a certain state of stress during the test.

In the existing studies on physiological stress, only a small number of sample data are used to carry out statistical analysis on individual data and extract some features obviously visible to the naked eye. Such research results may not be universal. In this work, we mainly focus on physiological changes under the influence of specific stressors, hoping to find some common physiological features from a large number of individuals, and use these features to train machine learning and deep learning models for people's physiological stress detection. In the future study, hope that can be used to identify the stress found some abnormal phenomenon, intervene to individuals, prevent some adverse events occur (such as suicide), through useful guidance, ease the stress on abnormal individuals.

To contribute to some existing research problems, we designed a data acquisition program and stress induction test. Based on previous studies, the Kraepelin test and Stroop test and their derivative tests were selected as stress-induced stimulus sources. The experiment was divided into eight sessions. The first session did nothing and was used as the baseline reference test, and the next seven sessions are run through the task-related tests we set. After the end of each test, the subject would report the current test's subjective stress as a reference for our subsequent stress detection and rating. At the same time of each session, we collect PPG, EDA data, and record facial video with an ordinary webcam, which could be used to extract facial information and eye gaze trajectory. We recruited a total of 120 participants to participate in our experiment, and each recorded eight pieces of data on each channel(about 10 min in total). A summary of the dataset is shown in Table 1. These data are used to analyze the correlation between physiological signals and stress, and the machine learning method is used for stress detection as a benchmark on this dataset.

Table 1. Summary of the MDPSD dataset.

Number of participants	120
Stressors	The Stroop Color-Word Test, the Rotation Letter Test, the Stroop Number-Size Test, and the Kraepelin Test
Data length	20s (the baseline stage), 150s (the Kraepelin Test), and 60s (all the rest of tests)
Stress ratings	0–4, indicating the change from feeling no stress to feeling Very high
Data types	EDA, PPG, and Facial videos

2   Related Work

In the circumplex model [30], stress also is an affective state, so stress is often studied as an affective state in conjunction with other emotions. This section introduces the research work based on physiological signal-related and lists emotion or stress datasets presented in the study.

2.1   Stress Detection

In 1998, Picard et al. [16] developed a system to quantify relevant physiological features under emotional stress, collecting four different physiological data of skin conductance, response, muscle activity, and heart activity of car drivers. The physiological data is collected and processed by sensors worn on the driver and the onboard portable processor. The driver's physiological signal has changed significantly when controlling the speed on the highway. They believe that stress and its influencing factors should be modeled. The relevant algorithm research results can be applied to car drivers to determine whether the driver is under pressure and intervene in the driver through cell phone calls or navigation assistance [15].

Healey and Picard are pioneers in stress detection, discovering that stress can be detected by physiological sensors [17]. Researchers conducted various studies using a combination of signal processing and machine learning (ML) for stress detection. Most of the data are from respiratory sensors [17,19], electrocardiogram sensors [3,29], heart rate (HR) sensors [34], acceleration (ACC) sensors [11,31,33], electrodermal electrical activity (EDA) sensors [20,29,35], blood volume pulse (BVP) sensors [14], electroencephalograph (EEG) sensors [4,28] and electromyography (EMG) sensors [40].

In [32], a mobile ECG sensor was used to collect physiological data during the Stroop test [37], and the pressure was detected by analyzing the ultra-short-term heart rate variability. The results showed that the ultra-short-term features could be used for stress monitoring, but they did not use any machine learning classification tools to support this conclusion. Lee et al. [25] used EDA signals and relevant characteristics of skin temperature to distinguish stress and non-stress states. The study in [41] combined EDA signals, blood volume pulse, pupil diameter, and skin temperature to measure the pressure of participants taking the altered Stroop test. Amandeep et al.[9] proposed a new method for psychological stress detection based on the nonlinear entropy feature extracted by empirical mode decomposition.

Gjoreski et al. [13] analyzed the problem of stress detection using off-the-shelf wrist devices with biosensors in laboratory conditions, and applied the extracted laboratory knowledge to real-life data collected entirely in the field. Their methods are applied to actual data, and some results are obtained. A total of 55 days of real data were collected from 5 subjects. The model detected 70% stress events with an accuracy of 95%. However, this experiment was conducted on only 5 subjects, and all data in this study were collected using the Empatica device. Hence, the proposed context-based approach is biased toward this device.

2.2   Datasets

The Driver stress dataset [17] includes physiological data of 24 participants (ECG, EDA, EMG, respiration). It recorded data on one resting condition and two driving missions (city streets and freeways near Boston, Massachusetts). Depending on the traffic conditions, the two driving tasks can last between 50 and 90 min. Through questionnaires and scores from observable events, three study conditions (rest, highway, and city) were mapped to low, medium, and high-stress levels. Therefore, this data set is helpful for the development of real-life stress monitoring methods. However, one limitation of the dataset is that it is obtained at a low sampling rate (for example, EMG 15.5 Hz).

Non-EEG [8] is a dataset containing physiological data (EDA, HR, skin temperature, arterial oxygen level, and 3-axes acceleration) of 20 subjects (4 female). The dataset recorded three different stress conditions (physical, cognitive, and emotional) and relaxation tasks. Having subjects jog on a treadmill at 3 miles per hour can cause physical stress. To trigger cognitive stress, the subjects counted backward from 2,485 at intervals of minus 7. Finally, watching a zombie apocalypse movie clip causes emotional stress. This dataset is particularly interesting because it contains only wearables based data. The low sampling rates (1 Hz and 8 Hz) of the devices used in the non-EEG dataset are a significant limitation. In addition, HRV information cannot be retrieved because no EEG or PPG data have been recorded, and this parameter is associated with stress recognition in previous work such as Kreibig [23].

WESAD [35] is a dataset for wearable stress and emotion recognition, which contains data from the subject's emotional and stress stimulus experiment. It is a publicly available data set and includes data from 15 subjects (3 female) recorded in a laboratory setting. Each subject experienced three conditions: baseline (neutral reading task), amusement (watching some exciting video clips), and stress (the Trier Social Stress Test was used). WESAD features physiological and movement data, recorded from the wrist and chest-worn devices. It uses sensors to record data in various modes, including ECG, PPG, EDA, EMG, respiration, skin temperature, and 3-axes acceleration (ACC). Besides, the chest adjustment device has a high sampling rate (700 Hz). Overall, WESAD is a suitable dataset for emotion and stress recognition based on physiological signals.

3   Data Collection

3.1   Stress Inducing Methods

In most of the literature available to us, stress is generated by stimulating subjects in a controlled environment. Therefore, in order to simulate the physiological response under stress, we need to choose appropriate stressors suitable for laboratory use and apply these stressors to subjects and collect various physiological data under some stress state. This section introduces some of the commonly used stress stimulation methods in the study and some methods we have designed based on existing tests.

The International Affective Picture System (IAPS) [24] is widely used in sentiment research and consists of a set of pictures with emotions that are evaluated using 9 levels of arousal and valence dimensions. In [5,12,21], IAPS is also used for stress recognition. The Stroop Color-Word Test (SCWT) [37] is a task that requires the name of a series of words about colors. These words representing colors are displayed in consistent or inconsistent colors. It has been validated in physiologically measured responses [38] and is considered a reliable source of stress. Many studies use this method as a source of stress induction, such as [2,18,39]. Mental Arithmetic (MA) tests are thought to cause stress and have been used in stress studies [7,26,27]. There are other ways of stress stimulus that may be ethically controversial [36]. In this paper, we used four stressors in the data collection test: the classic Stroop Color-Word Test (SCWT), the Rotation Letter Test, the Stroop Number-Size Test (SNST) that we changed it in the thought of SCWT, and the Kraepelin Test (Fig. 1).

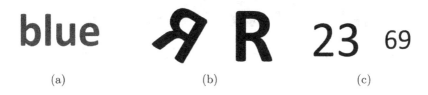

(a)   (b)   (c)

Fig. 1. Sample display of stressors used in the test. (a) Stroop Colour-Word Test (SCWT). (b) Rotation Letter Test. (c) Stroop Number-Size Test (SNST).

Stroop Color-Word Test (SCWT). We have selected four colors of red, green, blue, and yellow and their corresponding word representations. During the test, the words and colors are displayed randomly and randomly. Press the key to submit the judgment result (the left key indicates that the display color and word meaning are consistent, and the right key indicates that they are inconsistent), and each test displays one by one. This test has a certain amount of stress on people who are not color sensitive and can effectively stimulate them.

Rotation Letter Test. We selected four asymmetric uppercase English letters ('F', 'G', 'R', and 'Q') and made them into a specific size picture. During the test, two letter-images are displayed horizontally. Among them, the right letter image is randomly selected from the four we made in the front. The left letter image is obtained by center rotation or flipping and center rotation of the right image. The rotation angle is between 30 and 330° degrees and is an integer multiple of 30. The subject needs to determine what kind of transformation the left image is from the original image on the right, and also submit the judgment result through the left and right arrow keys of the keyboard (the left key represents flip and rotation, and the right button represents only rotation). This test can stimulate people who are not good at spatial imagination and can generate a heavy cognitive load. The test we did see in the literature have not used to do stress tests, and we should be the first to use this approach.

Fig. 2. Participant with sensors (PPG, EDA) during data collection.

Stroop Number-Size Test (SNST). This test is a variant of SCWT. It replaces the original colors and words with digital font size and numerical value size. Two numbers are printed on the screen in different font sizes. There are two possible situations. The value of one number and the font it displays are larger than the value and font of the other number. This situation is called consistent, and the other situation is not consistent. Similarly, the left arrow key on the keyboard represents consistency, and the right arrow key represents inconsistency.

Kraepelin Test. This test was proposed by German psychiatrist Emil Kraepelin in 1902 [22]. Like SCWT, it was initially used in psychological tests and later used in psychological stress tests. In this work, the subject needs to use mental arithmetic to calculate multiple consecutive addition questions containing 10 digits (range from 1 to 9). At the beginning of each question, the screen displays the calculation time in a conspicuous form. It is adding extra stress stimulus while highly focused on doing arithmetic.

For the first three tests, we also set a corresponding inversion test that added the cognitive process. Subjects were required to choose the wrong answer to submit based on the original regular judgment of the answer. By increasing the complexity of the task, the cognitive load and the sense of stress were increased.

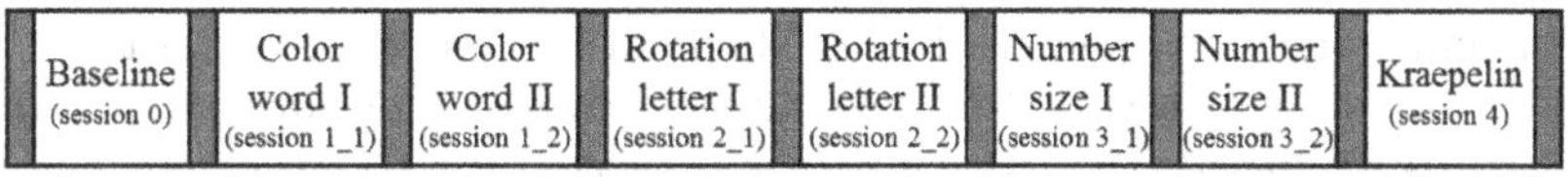

Baseline (session 0)	Color word I (session 1_1)	Color word II (session 1_2)	Rotation letter I (session 2_1)	Rotation letter II (session 2_2)	Number size I (session 3_1)	Number size II (session 3_2)	Kraepelin (session 4)

Fig. 3. The Lab study procedure. Green block includes rest time and the subject's self-assessment of current stress, II is inversion test of I. (Color figure online)

Table 2. Parameters of customized physiological sensor device.

Power supply	USB
Sampling frequency	200 Hz
Baud rate	57,600 Bd
EDA range	100–2,500 kΩ
EDA acquisition	External current excitation
PPG acquisition	Mechanism changes in blood volume of the blood vessels at the end of the finger

3.2 Expremental Setup

Materials. The experiments were performed in a closed room with lighting control. A computer used in the experiment was placed in the room for volunteers to complete a series of set stress testing experiments, and at the same time, used to collect various data during the experiment. Digital camera, EDA, and PPG signal acquisition equipment were connected with PC via USB port. The stress test and data collection programs are written in python programming. While the volunteers are completing the tests, the program collected facial video data, EDA, and PPG data simultaneously. The sensor device parameters are shown in Table 2. The test used a monitor to demonstrate stress test experiments. Each test requires the subject to use the left and right arrow keys of the keyboard or the mouse to enter the answer to complete each test, and each test takes between 1 min and 2 min and 30 s.

Protocol. The experiment was approved by the local ethics committee and performed under the ethical standards, as laid down in the Declaration of Helsinki. Individuals with a medical history that could affect the physiological system under study were excluded. Some participants have previously participated in another data acquisition test experiment in the laboratory, which can better adapt to the experimental environment. The data collection is divided into eight chips, includes the relaxation baseline test, the Stroop color-word test, the Stroop color-word reversal test, the rotation letter test, the rotation letter reversal test, the Stroop number-size test, the Stroop numbers-size reversal test, and the Kraepelin test. For each subject, the order of 8 sessions was the same. The test process is shown in Fig. 3. The baseline session is 20 s, and the next 6 sessions of the experimental collection time are 1 min. The final Kraepelin test collected 2 min and 30 s. At the end of each test, the subject gave his/her subjective psychological stress feeling for the current test and relaxed for about 10 min. Let the subjects return to a calm and relaxed state without being tired and distracted by long clicks. Data was not collected and recorded during the rest period. Figure 2 shows the scenario of data acquisition.

Subject Preparation. 120 undergraduates (72 males, mean age = 22 years old) participated in this experiment. To avoid the influence of some additional factors on the experiment process, the subjects were asked to turn off the mobile phone and other communication devices after entering the experiment room. Subjects were calm and relaxed to sit down and rest for a while, begin to understand the experiment process and points for attention, and were told the laboratory equipment is harmless to eliminate their psychological fear. After the procedure was fully explained, the participant signed an informed consent form approved by the local ethics committee to collect data during the experiment. Subjects after sitting before the PC display used in the experiment, adjust the seat's position, make sure that the experiment process, is placed on display at the top of the camera (the camera was calibrated) can accurately record facial information acquisition. Two finger-clips of the EDA sensor were placed in the left index finger and middle finger in a fixed order, and the finger clip of the PPG sensor was placed in the left nameless finger. After wearing, the left hand was placed on the table in a form that subjects felt comfortable, and the subjects were required not to move during each experiment.

Subjects were not required to do anything during the relaxation phase and were required to complete the test following the test rules mentioned in Sect. 3.1. Each test consists of several questions that appear one after the other. Subjects are asked to determine the answer as quickly as possible after each question appears on the screen, and then submit it. If the answer is wrong, there is an error tone when submit. Record the answer time for each test question. The number of right and wrong questions for each session of the test is also recorded. According to the experimental requirements, these simple data can be used to make a preliminary judgment on whether the subject seriously participates in the experiment. After a subject has completed all tests, the experimenter judges whether the data is valid by checking the number of wrongs and the time spent on the questions. At the end of each test, the subject should give the current test's subjective feedback, briefly describe the feelings during the test, and feedback a stress status value according to the test experience. The stress label is five discrete values from 0 to 4, indicating the change from feeling no stress to feeling very high.

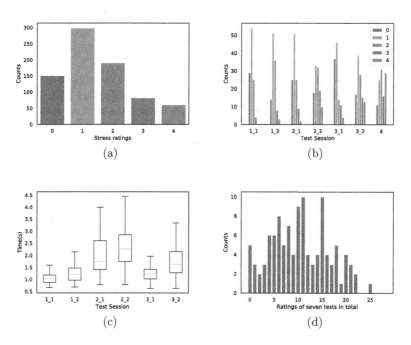

Fig. 4. (a) Distribution of stress level. (b) Ratings of each session. (c) Time used for each session. (d) Ratings of each participant.

4   Rating Analysis

As mentioned above, one of our goals in collecting such a database on stress was to examine whether stimuli from different stressors effectively stimulated subjects' stress state. In this section, we compare the stress arousal ratings of subjects under different stressors. Self-assessment is a conscious reflection of the user's stress perception when participating in different stress test stimuli. Any differences in ratings for stressors also affect the recognition of physiological responses.

Figure 4a shows the feedback distribution of stress levels for all different stress source tests, obtained by a total of 112 participants, each of whom had seven test sessions. The horizontal axis shows the stress level, and from left to right indicates no stress, slight stress, general stress, high stress, and very high stress.

As can be seen from the figure, among all the test results, 19.26% of the test process was considered to have no stimulation to produce any feeling of stress. About a quarter of the tests, the subjects felt stressed, but they felt weak. Less than a tenth of the test participants thought it could cause them to have a strong feeling of stress. Generally speaking, on the one hand, the selected stressors in the experiment should be reasonable, which can stimulate their different levels of stress feeling and facilitate us to analyze the changes of physiological signals in their corresponding states. On the other hand, the overall stress arousal rating is more in 1 and 2, which will not produce intense stimulation to the participants and minimize their normal life impact.

Figure 4b shows the distribution of arousal ratings for each stressor. From left to right, their average stress arousal ratings are 1.04, 1.42, 1.21, 1.73, 1.10, 1.73, and 2.25. Almost half of the first three groups found the test to be somewhat stressful, but very mild, and no or almost no one found the test to be very stressful. In the fourth and final tests, there was a significant increase in the number of participants reporting the highest levels of stress, particularly in the final group. About a quarter of the participants rated the highest levels of stress. In the first six groups, two adjacent groups can form control. The former is a routine test. The latter is based on the former added cognitive process, and the latter group's overall arousal score is 0.5 higher than the former group on average. On the whole, the distribution of each group of tests is not the same. For different participants, different feelings of pressure are different. Almost every group of tests can stimulate the subjects' sense of stress from a specific aspect. Figure 4c shows the average time taken by each participant to complete each test. The boxplot of the first six groups of tests is shown(because the first six groups are instantaneous reaction judgment tests for participants, which have a certain similarity, while the last continuous addition requires a long time and effort of participants, and there is no comparability). In comparing the two groups, it was clear that the increase in cognitive processing led to a longer overall time per question. The middle rotation letter test takes more time to complete the two sets of tests because it involves transforming the graph space and because the two alternative answers to the test are not opposed to each other.We conducted a statistical analysis of each subject's overall stress arousal score that all the seven tests were counted. Theoretically, the stress arousal value could be anywhere from 0 to 28. The actual statistical results are shown in Fig. 4d. Several participants reported that none of the tests had caused them any stress, and a small number had a slight feeling of stress overall. Most participants reported stress arousal in a moderate range, with none of them achieving the highest scores.

In this section, we demonstrate the stress arousal of all participants from multiple perspectives. In general, it shows the wakefulness of the 5-value interval, which should indicate that the stimulus we choose is reasonable to some extent, and it is reflected in each arousal, which is helpful for us to analyze the corresponding physiological signals. In different test experiments, stress arousal presents different distributions, indicating a certain degree of differentiation between different test experiments. Cognitive load is given to participants from different perspectives, and subjects are stimulated to different degrees. By combining the average time consumed per question in each section of the test with the stress rating reported by the subjects, the longer the response time required, the higher the overall stress arousal, which is reasonable from the perspective of cognitive load. The test of increasing the cognitive processing process can better stimulate the stress generation of the subjects. The continuous addition test requires the subjects to be highly focused for an extended period without any relaxation, which dramatically stimulates their stress.

5   Data Process

5.1   PPG Features

First, we use the fast Fourier transform to obtain the frequency distribution of the PPG signal. The result is shown in Fig. 5, and the useful frequency range of the pulse is 25 Hz. Use a bandpass filter to filter out high-frequency interference and retain the signal frequency range 0 Hz and 25 Hz. Since the individual physiological signals are very different, usually different people have different physiological signals, even if the same person is different at different times and in different environments. In the study, the subjects are composed of many different individuals, so it is necessary to remove the differences among individuals, reduce the errors caused by individual differences, and get the changes of physiological signals' interior features with different emotional states. The process of removing the difference is that the signal value in each subject's stress state is subtracted from its mean value in the baseline state, to obtain a physiological signal to remove the difference. Calculate the time and frequency domain features of the signal.

$$s(t) = X(t) - \overline{X}_{baseline}, \tag{1}$$

where $s(t)$ is the physiological signal after removing the baseline difference, $X(t)$ is original signal of test session, and $\overline{X}_{baseline}$ is mean of baseline session.

Calculate the first-order difference and second-order difference of signal data in the time domain. The first-order difference reflects the changing trend and speed of the signal. If the absolute value of the first-order difference is taken, the changing trend is ignored, and only the speed of change is considered. The first-order difference can be used to detect local extreme points of the signal, and the second-order difference can be used to detect local inflection points of the signal. Normalized signal data, first-order difference data, and second-order difference data respectively calculate the following features: Mean, median, standard deviation, maximum, minimum, difference between maximum and minimum, the ratio of the number of minima to the total number, the ratio of the number of maxima to the total number. Mean value:

$$u = \frac{1}{N} \sum_{t=1}^{N} s(t), \tag{2}$$

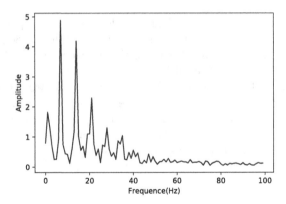

Fig. 5. Frequency distribution of PPG signal.

where N is the length of signal, s is a data sequence. Standard deviation:

$$\sigma = \{\frac{1}{N-1}\sum_{t=1}^{N}[s(t)-u]^2\}^{1/2}. \tag{3}$$

The first-order difference:

$$d(t) = s(t+1) - s(t). \tag{4}$$

Minimum ratio:

$$min_{Ratio} = \frac{Min}{N}, \tag{5}$$

where Min is the number of minima. Maximum ratio:

$$max_{Ratio} = \frac{Max}{N}. \tag{6}$$

The pulse is caused by the heartbeat, so the pulse and the ECG correlate and have apparent periodic changes. However, the periodic characteristics of the change are slightly delayed in time from the ECG. The relationship of typical pulse and ECG is shown in Fig. 6. In the figure, each character position of a pulse cycle waveform is marked: pulse initial point A, main wave peak P, repetitive wave peak T, and valley V between P and T. Detecting the specific positions in PPG signal is the premise of other operations, here we use the extremum detection method: according to the normal range of a pulse cycle, we set a search interval to detect the maximum value, the largest of the maximum values is the main wave peak P. Then P is used as a reference position to detect several other feature positions on both sides. The feature position in each cycle of the signal is sequentially detected. Here, because the high-frequency part of the signal is filtered out, it becomes smoother, and it is easy to locate the extreme point accurately and find the required feature point. Then we can get the time domain features through these positions:

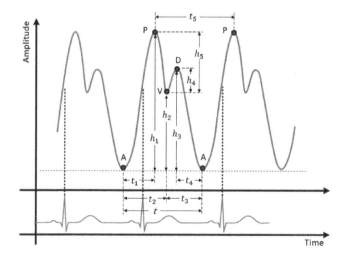

Fig. 6. The relationship between PPG and ECG.

PPG time interval: time to reach peak P (t1), duration of the main wave (t2), duration of repetitive wave (t3), the interval of relapse of repetitive wave (t4), interval time between two adjacent peaks P (t5) and the time t of a fluctuation cycle, the ratio of the main wave rise time and the pulse cycle, the time ratio of the main wave and the repetitive wave, the time ratio of the main wave and the repetitive cycle, and the time ratio of the repetitive wave and the repetitive cycle. Record data for each test segment and calculate the mean and standard deviation of the above features.

PPG amplitude: height of main wave crest (h1), the height of trough V (h2), the height of dicrotic wave relative to initial point (h3), the height of crest D relative to trough V (h4), main wave crest and repetitive wave peak amplitude ratio, main wave peak and trough V height ratio. The mean and standard deviation of the above characteristics are also calculated for the data collected in each segment.

PPG time interval and amplitude: the ratio of the amplitude of the peak P to the time, the ratio of the height of the falling edge of the peak P to the time, the ratio of the amplitude of the rising edge of the peak D to the time, the amplitude of the falling edge of the peak D and the ratio of time.

For frequency-domain features, only using time-domain feature parameters to characterize PPG signals' information under different stress states is not comprehensive enough, and the frequency domain feature vectors of the signal also need to be considered. Generally speaking, the frequency domain feature vectors of the original signal are obtained by fourier transform. Therefore, the original signal, that is, the time domain signal is usually obtained by Fourier transform to obtain the corresponding signal frequency domain spectrum. Then calculate the frequency mean, median, standard deviation, maximum value, minimum value, the maximum and minimum difference to get six frequency domain features.

5.2 EDA Features

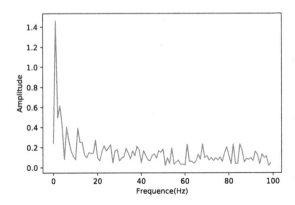

Fig. 7. Frequency distribution of EDA signal.

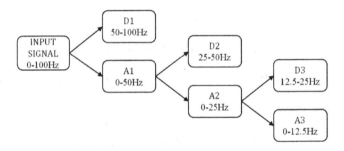

Fig. 8. An example of 3-layer wavelet decomposition. D represents the detail component, A represents the approximate component.

EDA signals can be used to study sympathetic nerve activity, thereby reflecting the individual's physiological and psychological activities and an evaluation index for emotional arousal. Unlike PPG, which is related to the heartbeat cycle, EDA's periodicity is manifested in different periods of the day, such as the sensitivity of the electrical response of the skin at night to rest and during the day. The stimulation will change, and the stimulation will produce adaptability after slowly, back to normal levels. Within a few minutes of the simulation experiment, EDA does not have noticeable periodic changes like PPG, so the time-domain features here are only mechanical statistical features, and no features closely related to physiological signal features are extracted. Among the time-domain features, 24 time-domain features such as the mean, median, standard deviation, first-order differential minimum ratio, second-order differential minimum value, and second-order differential maximum ratio are extracted.

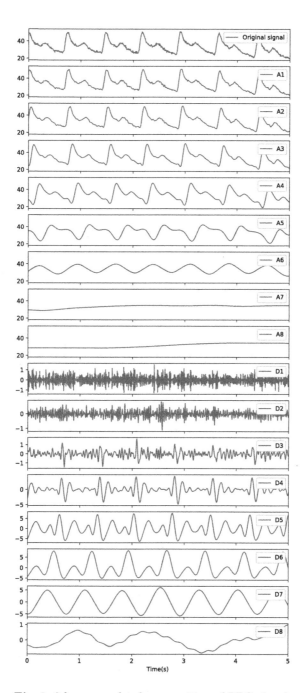

Fig. 9. 8-layer wavelet decomposition of PPG signal.

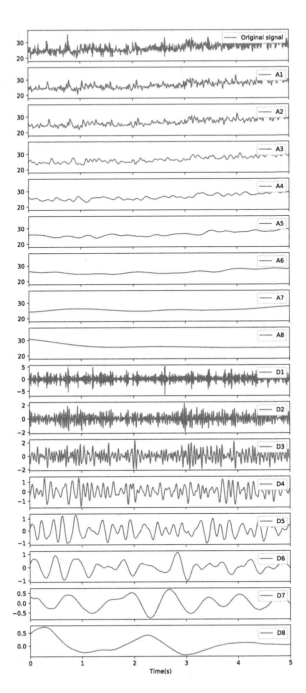

Fig. 10. 8-layer wavelet decomposition of EDA signal.

Like the PPG, first, use a low-pass filter to filter out glitches in the high-frequency part of the signal and keep the low-frequency part 5 Hz (see Fig. 7 for EDA's frequency distribution). Also, try to eliminate the differences in the individual's conductivity levels. Each piece of test data minus the calm state's baseline average is too low to obtain the data signal for subsequent use. In order to extract the frequency-domain features of EDA signals, first perform a discrete Fourier transform on the EDA data, and then calculate the frequency mean, median, standard deviation, maximum, minimum, maximum and minimum differences to obtain 6 frequency-domain features.

Wavelet transform is an excellent tool to extract the main shape parameters of the signal. Compared with the Fourier transform, the wavelet transform can describe both frequency and time domain information. Wavelet transform is a time-frequency localized signal analysis method with fixed window size but changeable shape (both time window and frequency window can be changed). Wavelet transform has higher frequency resolution and lower time resolution in the low-frequency part. It has a higher time resolution and lower frequency resolution in the high-frequency part, so it has the adaptability to the signal suitable for the analysis of physiological signals. The wavelet transform of signal $f(t)$ is defined as follows:

$$WT_f(a, \tau) = \frac{1}{\sqrt{a}} \int_R f(t) \psi \left(\frac{t - \tau}{a} \right) dt, \tag{7}$$

where $WT_f(a, \tau)$ is is called the wavelet transform coefficient, a is the scale factor, τ is the translation factor, and $\psi(\cdot)$ is the basic wavelet or wavelet mother function.

Wavelet transform decomposes the signal to different scales through low-pass filtering and high-pass filtering, which corresponds to different frequency bands. After high-pass filtering, the output coefficient is the detail coefficient; the coefficients are approximate coefficients after low-pass filtering. The approximate coefficients can continue to be high-pass filtering and low-pass filtering, decompose to the next level, and so on. Figure 8 is a schematic diagram of the signal 3-layer wavelet decomposition. The wavelet basis function selected in this paper is the db5 wavelet basis function in the Daubechies series of wavelets, and stationary wavelet transform is used. Each layer of wavelet transform produces an approximation A and a detail value D and then uses the approximation of the previous layer to perform wavelet decomposition to obtain the approximation and detail values of the current layer, and so on. The sampling rate of our database is $F_s = 200$ Hz. According to Nyquist's sampling theorem, the maximum frequency of the signal F_{max} should satisfy $F_s \geq 2F_{max}$, so the maximum frequency value that can be decomposed 100 Hz. In this paper, we do an 8-layer wavelet decomposition of the original EDA signal. We select five sets of wavelet coefficients (detail components D3, D4, D5, D6, and approximate component A6) to calculate energy entropy.

$$E_i = \sum_{t=1}^{N} x_t^2, \tag{8}$$

where N is the length of signal.

$$P_i = \frac{E_i}{E},$$ (9)

where E is he sum of E_i. Energy entropy as follows:

$$H_{EN} = -\sum_{i=1}^{n} P_i \log P_i.$$ (10)

To provide training data for the neural network model, we performed 8-layer wavelet decomposition on both the original PPG signal and the EDA signal, and the wavelet function settings are as described above. A part of the obtained detail components and approximate components is selected and input to the model. Each component coefficient is shown in Fig. 9 and Fig. 10.

6   Expriments and Results

Using machine learning and deep learning methods to train physiological signal data, the stress level is classified into two levels, three levels, and five levels. In the dataset, we divide the pressure into five levels. For two classifications, we will combine label "0" and "1" into one category, and label "2", "3" and "4" into one category; The labels "1" and "2" are merged, and the labels "3" and "4" are merged in three classifications. For the unbalanced dataset, we appropriately repeat the category with a small amount of data and delete a part of the samples with a large amount of data, and shuffle the order during training to ensure that the data set is relatively balanced. All models are trained using 5-fold cross-validation.

6.1   Feature-Based Approach

Each piece of PPG data includes: 24 features from the normalized signal, first-order difference signal, and second-order difference signal, 20 features are extracted from the temporal characteristics of the pulse, 12 features are derived from the amplitude of the pulse, 8 features are amplitude and time related, 6 frequency-domain features, form a 70-dimensional feature vector. Each piece of EDA data includes 24 features from filtered and normalized signals, first-order difference signal and second-order difference signal, 6 frequency-domain features, one feature that comes from the energy entropy after wavelet decomposition, and a total of 31-dimensional feature vectors. The physiological signals collected have two lengths of 60 s and 150 s. When calculating the statistical features, all the data is divided into 10 s each segment, the first 10 s of data are discarded, and the rest are retained–a total of 6384 samples.

We use five machine learning classifiers to detect stress through the extracted features. The five classifiers are Decision Tree, Random Forest, AdaBoost, Linear discriminant analysis, and k-Nearest Neighbor. Since the entire data processing flow is implemented in Python, we use the Scikit-Learn package for the classifier described above. The AdaBoost classifier uses decision trees as the basic estimator. For each classification algorithm based on decision tree (DT, RF, AB), information gain is used to measure the quality of the split decision node, and the minimum number of samples required to split a node is set to 20. For two ensemble learners (RF and AB), the number of base estimators is set to 100. And k is set to 9 for kNN. The other parameters use the defaults in the Scikit-Learn package. The experimental results are shown in Table 3.

Table 3. Accuracy of machine learning classifiers. DT, RF, AB, LDA and kNN are Decision Tree, Random Forest, AdaBoost, Linear discriminant analysis and k-Nearest Neighbor, respectively.

		DT	RF	AB	LDA	kNN
EDA	2	66.34%	73.87%	74.58%	63.12%	64.03%
	3	50.14%	56.00%	58.30%	49.70%	52.31%
	5	30.88%	33.71%	32.35%	31.58%	30.94%
PPG	2	64.60%	75.74%	74.07%	63.72%	61.52%
	3	45.32%	55.47%	55.92%	46.08%	46.38%
	5	27.84%	32.47%	32.58%	26.79%	30.29%

Table 4. Accuracy of Neural Network models (%).

	EDA			PPG		
	2	3	5	2	3	5
CNN1D	81.34	58.16	34.95	82.60	59.22	36.54
LSTM	82.19	61.04	36.32	81.41	60.55	36.07

6.2 Neural Network-Based Approach

We built two simple deep learning models. A one-dimensional convolution model and an LSTM model, Keras with TensorFlow as the back-end implements our model. The input of the model is the wavelet coefficients mentioned above, the input data of PPG are D3, D4, D5, D6, and A6, and the input data of EDA are D4, D5, D6, D7, and A7. We use the PyWavelets package to do wavelet decomposition on the signal. The length of the data that needs to be decomposed must be a multiple of 2^{level} in length, $level$ is the number of decomposed layers. The original data length after division is 1024, and the beginning and end of each piece of data exist overlap. After decomposed, data is divided into 5 s per segment in a non-overlapping manner, and the part that is less than the length is discarded.

The convolution model is followed by four one-dimensional convolution layers, a Dropout layer, and a fully connected layer, each convolution layer is followed by a maximum pooling layer. The length of the convolution window is 16, 8, 4, 2, respectively. Each layer is provided with 16, 64, 128, and 128 convolution kernels, and strides are all 2. The LSTM model contains only one LSTM layer, a Dropout layer, and a fully connected layer. All models use a cross-entropy loss function. Results in Table 4.

6.3   Analysis

The stress classification results from 5 different machine learning models can be seen. The overall accuracy is higher than random guesses, indicating that our dataset's physiological signals can distinguish the stress state. For EDA, the AdaBoost method achieved the highest accuracy in the binary and three classifications, and Random Forest achieved the highest score in the five classifications. Decision trees, linear discriminant analysis, and k-Nearest Neighbor are generally worse than the other two methods, but they are better than randomly guessing the classification results. The reason why their three methods are weak in results may be that they have different ways of learning data features and have limitations. The five methods all performed poorly on the five classifications. It may be that during the data collection process, the subjects have different feelings about their stress, and the pressure feedback they give has deviations, which makes the model unable to distinguish each sample well.

For PPG, Random Forest achieved the highest accuracy of 75.74% in the binary classifications. The performance of each classifier is similar to that of EDA in terms of PPG. The results obtained by Random Forest and AdaBoost methods are superior to the other three methods. Comparing the EDA and PPG classification results, the best result of PPG is one percentage point higher than the best result of EDA in binary classifications. However, in three classifications, the best results of both are obtained by AdaBoost, but the accuracy of EDA is Higher than PPG. Similarly, EDA on five classifications is slightly better than the result of PPG. We extracted 31 features on each EDA sample, and 70 features on each PPG sample, which is more than double that of EDA. More data features can provide more information to the model, which is beneficial to dichotomy. However, for multiple classifications, such information may conflict with each other without a reasonable degree of differentiation, which causes difficulties for the model in learning useful feature information.

It can be seen in Table 2. The stress detection accuracy of the neural network model is higher than the traditional machine learning methods. Here we just built a simple network model to verify the data's validity and did not add other advanced feature learning techniques. The accuracy of the two models we use is not much different. In binary classification, the LSTM model has better results on the EDA data than the convolutional network model, but the PPG data results are opposite. The accuracy of LSTM on three classifications is slightly better than the convolutional network model. Both models use wavelet coefficients as input; LSTM is more suitable for extracting time series data characteristics. From Fig. 9 and Fig. 10, comparing the wavelet coefficient with the original signal, the change of the physiological signal can be perceived in the wavelet domain, especially in the detail coefficient. We select the appropriate coefficient components as training data based on the characteristics of wavelet decomposition and the frequency domain distribution of physiological signal data. In the experiment, we verified the validity of the data. In the subsequent research, in addition to the improvement of the model, we will use the facial information and the physiological signals of the two channels together as a multimodal data training model to improve the accuracy of stress detection.

Table 5. Comparison with previous work in binary and three classification. This table compared the results of the EDA experiment in this article with the results of the EDA data collected using chest-based device in the WESAD dataset. DT, RF, AB, LDA and kNN are Decision Tree, Random Forest, AdaBoost, Linear discriminant analysis and k-Nearest Neighbor, respectively.

		DT	RF	AB	LDA	kNN
2	[35]	76.21%	76.29%	79.71%	78.08%	73.13%
	Our	66.34%	73.87%	74.58%	63.12%	64.03%
3	[35]	48.49%	45.00%	54.06%	67.07%	40.03%
	Our	**50.14%**	**56.00%**	**58.30%**	49.70%	**52.31%**

Schmidt et al. proposed the WESAD dataset, which has recorded data using two different devices (one based on the chest and one based on the wrist), including high-resolution physiological and motor modalities. Table 5 shows the accuracy of EDA signals in our dataset and the WESAD dataset to extract features and learning under different machine learning methods. It can be seen from the results that under the binary classification, the accuracy of the five machine learning models on our data set is slightly lower than that of the WESAD data. However, in the case of three-classification, our dataset performs worse than WEASD dataset only in LDA, and better in the other four methods. As can be seen from Table 3, our dataset's performance under the LDA method is generally worse than that of other methods. WESAD contains the data of 15 subjects, while our data set contains the data of 120 subjects. Our data is more generalized and performs better in the decision-based learning method.

7   Conclusion and Discussion

In response to some existing research problems, we have designed a set of data collection and stress induction test programs. The Kraepelin test, Stroop test, and its derivative tests were selected as stress-induced stimulation sources based on previous studies. Collected facial videos, PPG, and EDA data of 120 participants. These data are used to analyze the correlation between physiological signals and pressure and use machine learning methods for stress detection as the benchmark for this dataset. We also achieved better stress detection accuracy than the benchmark on simple neural network models. In the future, we hope to find some abnormal phenomena by identifying stress, intervene in individuals, prevent some adverse events (such as suicide), and reduce the stress of abnormal individuals through useful guidance.

In this article, we only verify the effectiveness of PPG and EDA for stress detection. In the next research, we need to extract useful information from the facial videos and combine them with two physiological signal channels to improve stress detection accuracy and robustness.

References

1. Acerbi, G., et al.: A wearable system for stress detection through physiological data analysis. In: Cavallo, F., Marletta, V., Monteriù, A., Siciliano, P. (eds.) ForItAAL 2016. LNEE, vol. 426, pp. 31–50. Springer, Cham (2017). https://doi.org/10.1007/978-3-319-54283-6_3
2. Alonso, J., Romero, S., Ballester, M., Antonijoan, R., Mañanas, M.: Stress assessment based on EEG univariate features and functional connectivity measures. Physiol. Measur. **36**(7), 1351 (2015)
3. Anusha, A., Jose, J., Preejith, S., Jayaraj, J., Mohanasankar, S.: Physiological signal based work stress detection using unobtrusive sensors. Biomed. Phys. Eng. Express **4**(6), 065001 (2018)
4. Asif, A., Majid, M., Anwar, S.M.: Human stress classification using EEG signals in response to music tracks. Comput. Biol. Med. **107**, 182–196 (2019)
5. Baltaci, S., Gokcay, D.: Stress detection in human-computer interaction: fusion of pupil dilation and facial temperature features. Int. J. Hum. Comput. Interact. **32**(12), 956–966 (2016)
6. Beatty, J., Lucero-Wagoner, B., et al.: The pupillary system. In: Handbook of Psychophysiology, vol. 2, pp. 142–162 (2000)
7. Bernardi, L., Wdowczyk-Szulc, J., Valenti, C., Castoldi, S., Passino, C., Spadacini, G., Sleight, P.: Effects of controlled breathing, mental activity and mental stress with or without verbalization on heart rate variability. J. Am. Coll. Cardiol. **35**(6), 1462–1469 (2000)
8. Birjandtalab, J., Cogan, D., Pouyan, M.B., Nourani, M.: A non-EEG biosignals dataset for assessment and visualization of neurological status. In: 2016 IEEE International Workshop on Signal Processing Systems (SiPS), pp. 110–114. IEEE (2016)
9. Cheema, A., Singh, M.: An application of phonocardiography signals for psychological stress detection using non-linear entropy based features in empirical mode decomposition domain. Appl. Soft Comput. **77**, 24–33 (2019)
10. Chrousos, G.P.: Stress and disorders of the stress system. Nat. Rev. Endocrinol. **5**(7), 374 (2009)
11. Fukazawa, Y., Ito, T., Okimura, T., Yamashita, Y., Maeda, T., Ota, J.: Predicting anxiety state using smartphone-based passive sensing. J. Biomed. Inform. **93**, 103151 (2019)

12. Giannakakis, G., et al.: Stress and anxiety detection using facial cues from videos. Biomed. Signal Process. Control **31**, 89–101 (2017)
13. Gjoreski, M., Luštrek, M., Gams, M., Gjoreski, H.: Monitoring stress with a wrist device using context. J. Biomed. Inform. **73**, 159–170 (2017)
14. Handouzi, W., Maaoui, C., Pruski, A., Moussaoui, A.: Short-term anxiety recognition from blood volume pulse signal. In: 2014 IEEE 11th International Multi-Conference on Systems, Signals & Devices (SSD14), pp. 1–6. IEEE (2014)
15. Healey, J., Picard, R.: SmartCar: detecting driver stress. In: Proceedings 15th International Conference on Pattern Recognition, ICPR-2000, vol. 4, pp. 218–221. IEEE (2000)
16. Healey, J., Seger, J., Picard, R.: Quantifying driver stress: developing a system for collecting and processing bio-metric signals in natural situations. Biomed. Sci. Instrum. **35**, 193–198 (1999)
17. Healey, J.A., Picard, R.W.: Detecting stress during real-world driving tasks using physiological sensors. IEEE Trans. Intell. Transp. Syst. **6**(2), 156–166 (2005)
18. Hou, X., Liu, Y., Sourina, O., Tan, Y.R.E., Wang, L., Mueller-Wittig, W.: EEG based stress monitoring. In: 2015 IEEE International Conference on Systems, Man, and Cybernetics, pp. 3110–3115. IEEE (2015)
19. Hovsepian, K., al'Absi, M., Ertin, E., Kamarck, T., Nakajima, M., Kumar, S.: cStress: towards a gold standard for continuous stress assessment in the mobile environment. In: Proceedings of the 2015 ACM International Joint Conference on Pervasive and Ubiquitous Computing, pp. 493–504. ACM (2015)
20. Huysmans, D., et al.: Unsupervised learning for mental stress detection-exploration of self-organizing maps. Proc. Biosignals **2018**(4), 26–35 (2018)
21. Khalilzadeh, M.A., Homam, S.M., HOSSEINI, S.A., Niazmand, V.: Qualitative and quantitative evaluation of brain activity in emotional stress (2010)
22. Kraepelin, E.: Die arbeitscurve, vol. 19. Wilhelm Engelmann (1902)
23. Kreibig, S.D.: Autonomic nervous system activity in emotion: a review. Biol. Psychol. **84**(3), 394–421 (2010)
24. Lang, P.J., Bradley, M.M., Cuthbert, B.N.: International affective picture system (IAPS): technical manual and affective ratings. NIMH Center for the Study of Emotion and Attention, vol. 1, pp. 39–58 (1997)
25. Lee, M.h., Yang, G., Lee, H.K., Bang, S.: Development stress monitoring system based on personal digital assistant (PDA). In: The 26th Annual International Conference of the IEEE Engineering in Medicine and Biology Society, vol. 1, pp. 2364–2367. IEEE (2004)
26. Lundberg, U., et al.: Psychophysiological stress and EMG activity of the trapezius muscle. Int. J. Behav. Med. **1**(4), 354–370 (1994)
27. McDuff, D., Gontarek, S., Picard, R.: Remote measurement of cognitive stress via heart rate variability. In: 2014 36th Annual International Conference of the IEEE Engineering in Medicine and Biology Society, pp. 2957–2960. IEEE (2014)
28. Minguillon, J., Perez, E., Lopez-Gordo, M.A., Pelayo, F., Sanchez-Carrion, M.J.: Portable system for real-time detection of stress level. Sensors **18**(8), 2504 (2018)
29. Muaremi, A., Bexheti, A., Gravenhorst, F., Arnrich, B., Tröster, G.: Monitoring the impact of stress on the sleep patterns of pilgrims using wearable sensors. In: IEEE-EMBS International Conference on Biomedical and Health Informatics (BHI), pp. 185–188. IEEE (2014)
30. Posner, J., Russell, J.A., Peterson, B.S.: The circumplex model of affect: an integrative approach to affective neuroscience, cognitive development, and psychopathology. Dev. Psychopathol. **17**(3), 715–734 (2005)

31. Ramos, J., Hong, J.H., Dey, A.K.: Stress recognition-a step outside the lab. In: PhyCS, pp. 107–118 (2014)
32. Salahuddin, L., Cho, J., Jeong, M.G., Kim, D.: Ultra short term analysis of heart rate variability for monitoring mental stress in mobile settings. In: 2007 29th Annual International Conference of the IEEE Engineering in Medicine and Biology Society, pp. 4656–4659. IEEE (2007)
33. Sano, A., Picard, R.W.: Stress recognition using wearable sensors and mobile phones. In: 2013 Humaine Association Conference on Affective Computing and Intelligent Interaction, pp. 671–676. IEEE (2013)
34. de Santos Sierra, A., Ávila, C.S., Casanova, J.G., Del Pozo, G.B.: Real-time stress detection by means of physiological signals. Recent Appl. Biometrics **58**, 4857–65 (2011)
35. Schmidt, P., Reiss, A., Duerichen, R., Marberger, C., Van Laerhoven, K.: Introducing WESAD, a multimodal dataset for wearable stress and affect detection. In: Proceedings of the 2018 on International Conference on Multimodal Interaction, pp. 400–408. ACM (2018)
36. Sharma, N., Gedeon, T.: Modeling a stress signal. Appl. Soft Comput. **14**, 53–61 (2014)
37. Stroop, J.R.: Studies of interference in serial verbal reactions. J. Exp. Psychol. **18**(6), 643 (1935)
38. Tulen, J., Moleman, P., Van Steenis, H., Boomsma, F.: Characterization of stress reactions to the Stroop color word test. Pharmacol. Biochem. Behav. **32**(1), 9–15 (1989)
39. Visnovcova, Z., et al.: Complexity and time asymmetry of heart rate variability are altered in acute mental stress. Physiol. Meas. **35**(7), 1319 (2014)
40. Wijsman, J., Grundlehner, B., Liu, H., Penders, J., Hermens, H.: Wearable physiological sensors reflect mental stress state in office-like situations. In: 2013 Humaine Association Conference on Affective Computing and Intelligent Interaction, pp. 600–605. IEEE (2013)
41. Zhai, J., Barreto, A.: Stress detection in computer users based on digital signal processing of noninvasive physiological variables. In: 2006 International Conference of the IEEE Engineering in Medicine and Biology Society, pp. 1355–1358. IEEE (2006)

Improving Small-Scale Dataset Classification Performance Through Weak-Label Samples Generated by InfoGAN

Meiyang Zhang[1], Qiguang Miao[2], Daohui Ge[2], and Zili Zhang[1,3](✉)

[1] College of Computer and Information Science, Southwest University,
Chongqing 400715, China
[2] School of Computer Science and Technology, Xidian University,
Xi'an 710071, China
[3] School of Information Technology, Deakin University, Locked Bag 20000,
Geelong, VIC 3220, Australia
zhangzl@swu.edu.cn

Abstract. Sufficient training data typically are required to train learning models. However, due to the expensive manual process for labeling a large number of samples, the amount of available training data is always limited (real data). Generative Adversarial Network (GAN) has good performance in generating artificial samples (generated data), the generated samples can be used as supplementary data to make up for the problem of small dataset with small sample size and insufficient diversity. Unfortunately, the generated data usually do not have annotation label. To make better use of the generated data, a learning framework WGSForest is proposed, which realizes the use of real data and generated data to train the classifier jointly.

In the WGSForest model, the supplementary data is generated through an improved InfoGAN to increase the amount and diversity of training data. Moreover, the generated supplementary data will be weakly labeled through InfoGAN. We utilize the advantage of deep forest on small dataset and take the generated data with a weak label as the supplement of real training data to optimize deep forest. In detail, The cascade forest in the improved deep forest (SForest) dynamically updates each generated data label to proper confidence, then the real data and generated data are combined to train the following layers of the improved cascade forest jointly. Experiment results showed that adding the weak label generated data effectively improves the classification performance of deep forest. On mnist (1000) subset, 100% generation rate can obtain 1.17% improvement, and 100% generation rate can obtain 6.2% improvement on 1000 cifar10 subset. Furthermore, each dataset can determine performance at a specific generation rate.

Keywords: Generated data · Deep forest · Small-scale data · InfoGAN

© Springer Nature Singapore Pte Ltd. 2021
H. Mei et al. (Eds.): BigData 2020, CCIS 1320, pp. 83–95, 2021.
https://doi.org/10.1007/978-981-16-0705-9_6

1 Introduction

Finding the corresponding labels for the data manually is costly. Moreover, many tasks failed to obtain a large amount of data due to personal privacy and data security. Many researches have been proposed to improve the learning ability of classifier in small-scale dataset. Among different classification tasks, image classification is a common but important task in the digital image area, and it is a typical computer vision problem that we usually require sufficient training data to learn a discriminative model. In [1], Facial features were learned from a small number of face images by combining distinctive facial areas. Then, combined images were incorporated with the original images to train the CNN network. A convolutional self-encoding neural network [2] was proposed to encode and decode original images to obtain new images. The difference between original images and new images was minimized to train the CNN parameters. Utkin [3] improved Siamese network through a 2-channel network, then the input is a pair of sample data, not a single sample. The network can be trained by judging the similarity of the two networks.

Generative Adversarial Network (GAN) was proposed to generate artificial samples with perceptual quality [4]. Since then, several improved approaches [5–7] were proposed to further improve the quality of generated samples. Current researches in GAN typically considered the high quality of the sample generation with and without semi-supervised learning in GAN vivo [8–10]. However, how to efficiently use the generated data in vitro is still an outstanding question. Our focus is to use generated data to directly participate in training as the supplementary of limited labeled data.

In all existing methods by using GAN in vitro, there are two main challenges in order to assure the better performance: 1) high quality and diversity data generated by GAN, and 2) a better strategy to integrate the generated data into the learning model [11,12]. Our focus is mainly on solving these two problems. First, the input to the generator is a continuous noise without any constraints, so the GAN cannot control the features of generated images by controlling certain dimensions of input. To this end, our model obtained the generated data through an Information Maximizing Generative Adversarial Networks (InfoGAN) [8], which adopts latent codes to vary image characteristic, thus it increases the amount and diversity of generated samples to a certain extent. Moreover, the generated data will be weakly labelled with classifier network of infoGAN. Second, current researches [11,12] generated unlabeled data via DCGAN [6]. The [11] assigned virtual labels to the generated data with a uniform label distribution over all the pre-defined training classes. However, during the actual GAN training process, only certain real data from some classes (not all pre-defined training classes) are used in GAN training at each iteration to generate artificial data following a continuous noise distribution [6]. Furthermore, it may not be correct to assign the same label to different generated data if the generated data have distinct visual differences. Based on the above drawbacks, MpRL [12] dynamically assigned a different virtual label to each generated data in every iteration. Different from the above methods, our work employs the SForest [13] to make the generated data label become more strong through the dynamical update of initial label. Then, the real data and generated data are combined to train the following layers of the deep forest jointly.

Deep forest [14] is a decision tree ensemble approach which is highly competitive with deep neural networks, and it can work well even when there are only small-scale training data. SForest was proposed to improve the classification performance of deep forest in small-scale datasets. To improve the deep forest generalization ability on small-scale data, the proposed WGSForest train the improved infoGAN on the real dataset and generate the supplement data through choosing categorical and continuous codes as infoGAN input. Meanwhile, The generated data quality is improved by considering the label loss information in the infoGAN discriminator training process, and the weak label of generated data is annotated by the Q network. Then, these weakly labeled generated data are fed into the SForest model to update the weak label by the trained SForest with real data. Once the label of generated data became strong, the generated data and the real data will be integrating to regularize the learning process of the classifier. The work reported in this paper includes the following three aspects:

(1) Sample generation. The improved InfoGAN generates high-quality data by considering the label loss of training data and vary the image style through latent code. Moreover, the generated data will be annotated with the weak label. Thus the relationship between the generated data and pre-defined training classes can be substantially built, which makes the generated data more informative when they incorporated with the real data in training. We found an average (+2.06%, +1.99%, 14.74%) improvement in (MNIST, ORL, Cifar10) dataset at 100% extra data rate.
(2) The joint training method. The cascade forest in SForest strengthens generated data label confidence, and then the generated data incorporated with real data in the following training process. We observed that the generated data as supplement data has 0.12% and 0.73% slight worst compared with real data as supplement data in MNIST and Cifar10 dataset.
(3) Experiments. Qualitative analyses are given to the proposed methods. Also, comprehensive quantitative evaluations are carried out to verify the performance of the proposed method on different small-scale datasets.

The rest of this paper is organized as following. We first review some related work in Sect. 2. Then, the implementation details of the proposed WGSForest model are provided in Sect. 3. The experiments are described in Sect. 4. The conclusion is in Sect. 5.

2   Related Work

2.1   Generative Adversarial Networks

Generative adversarial networks (GAN) learn two sub-networks: a generator and a discriminator. The discriminator reveals whether a sample is fake or real, while the generator produces samples to cheat the discriminator. The GAN was firstly proposed by Goodfellow et al. [4] to generate images and gain insights into neural networks. Then, DCGAN [6] provided some tricks to improve the stability of GAN training and avoid the model crash. Salimans et al. [10] achieved a state-of-art result in semi-supervised classification and improved the visual quality of GAN. InfoGAN [8] learned interpretable representations by introducing latent codes.

GAN also demonstrate potential in generating images for specific fields. Pathak et al. [15] proposed an encoder-decoder method for image inpainting, where GAN was used as the image generator. Reed [16] used GAN to realize the composite image through text information. Wu [17] learned the shape of 3D objects by using 3D-GAN. Unlike the sophisticated model that focuses on sample generation, we only use the infoGAN to generate samples with a weak label from the training data and show that these samples help improve discriminative learning.

2.2 Deep Forest

Deep forest [14] is an alternative model to deep learning structure. In contrast to deep neural networks, deep forest not only has much fewer hyper-parameters but also is easy to be theoretically analyzed. Meanwhile, the number of cascade layers can be adaptively determined such that the model complexity can be automatically set, enabling deep forest to perform excellently even on small-scale data. In classification improvement, Wen, Hong, et al. [18] took GBDT as classifier unit, and the cross-entropy of each leaf node was transferred to the next layer. Strong and weak correlation features were selected to boost feature importance. The dense connectivity was adopted to augment the data feature and SAMME.R boosting method was used to update the sample weight [19]. To increase efficiency, Pang, Ming, et al. [20] used sub-sampling in multi-grained scanning to decrease memory consumption and runtime significantly. The confidence screening mechanism filtered the high prediction confidence which directly transfers to the final layer. In small-scale data classification, Guo, Yang, et al. [21] adopted a multi-class-grained scanning strategy by using different training data to train different forest respectively, and the boosting strategy was used to emphasize more important features in cascade forest. Due to the small sample size and the class-imbalance in small-scale data, SForest [13] encouraged the diversity of ensemble and consider the augment features on each layer. The skip connection operation was injected into cascade forest and five different types of classifiers were used to improve the diversity of the ensemble and boost important features in forest learning.

2.3 SForest for Small-Scale Data Classification

Due to the small sample size and the class-imbalance in small-scale data, the SForest [13] employed multi-class-grained scanning strategy to encourage the ensemble diversity by using different class of training data to train different forest respectively, which can remove the risk of class-imbalance. From Fig. 2(a) train process illustrates, the raw training data is divided into specific sub-dataset by class label, so each sub-dataset has two types of samples (positive in this class, negative not in). Then each sub-dataset is used to train a forest solely to sense a specific class. Meanwhile, the out-of-bagging score is developed to measure the fitting quality of each forest.

The architecture of SForest is illustrated in Fig. 2, the skip connection is injected in cascade forest to augment the features of the next layer and five different types of classifiers (random forests, completely random forests, K-nearest neighbours, logistic regression and Xgboost) are used to improve the diversity of ensemble. In the multi-class-grained scanning, one sliding window contains two group forest ensembles. In the cascade forest, the transformed feature vectors, augmented with the class vector generated by the previous layer and the class vector is generated by the layer before the previous layer, will then be used to train the following cascade layers. Moreover, the top-k features are selected to extract the standard deviation boosting feature in each forest. This procedure will be repeated until convergence of validation performance (Fig. 1).

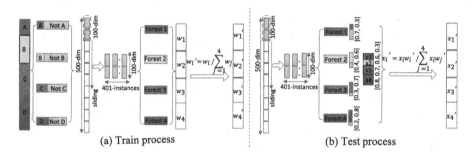

(a) Train process   (b) Test process

Fig. 1. Multi-class-grained scanning.

3   The Proposed WGSForest Model

The WGSForest includes two stages: sample generation and the joint training, we first introduce an improved InfoGAN network used for generating weakly labeled samples. Then we show the joint training method based on the SForest.

3.1   Samples Generation

The GAN uses a continuous input noise vector z, while imposing no restrictions on how the generator use this noise, causing the individual dimensions of z not correspond to semantic features of the data. InfoGAN is a generative adversarial network that also maximizes the mutual information between the latent variables and the observation. InfoGAN decomposed the input noise vector into two parts:

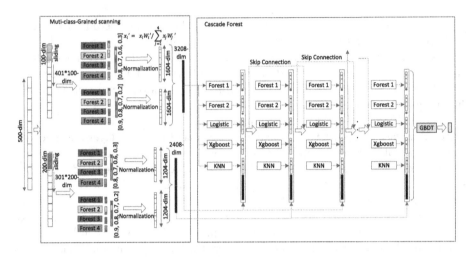

Fig. 2. Overall procedure of SForest.

(i) z, which is treated as the source of incompressible noise (ii) c, which is called the latent code and will target the ent structured semantic features of the data distribution. The generator G is trained to generate new sample $G(z, c)$. Then the correlation of c and $G(z, c)$ is expected to be higher, it means the mutual information term $I(c, G(z, c))$ between c and $G(z, c)$ should be larger. The discriminator D input real data x and fake data $G(z, c)$ and distinguish samples from the true data distribution P_{data} and the generators distribution P_G. Finally, classification network Q share all convolutional layers as D and there is one final fully connected layer to output parameters for the conditional distribution $Q(c|x)$. Therefore the infoGAN objective function is:

$$\min_G \max_D V_I(D, G) = V(D, G) - \lambda I(c; G(z, c)) \tag{1}$$

We consider the real data label in training the discriminator D, which can improve the performance of D and further improve the performance of G through the cyclic reaction. The objective function $V(D, G)$ part is converted to:

$$\min_G \max_D V(D, G) = E_{x \sim P_{dese}} [\log D(x)] + E_{x \sim P_G}[\log(1 - D(G(z))] \\ = E[\log P(C = c'|x = real) + E[\log(1 - P(C = \text{ fake } |x = \text{ fake }))] \tag{2}$$

where c' represents the class of input real sample x. The generated sample will be weakly labeled through the network Q. The supplement dataset from infoGAN can be obtained off-line and stored. Moreover, The amount of the generated supplementary data can be controlled by the generated data ratio u, which is computed by dividing the number of generated samples by the size of the original training set.

3.2   The Joint Training Method

InfoGAN generate the supplement dataset with the weak (initial) label. This weak label is the driver which determines the number of layer i in cascade forest for updating the generated sample label, so the initial label can be used without being accurate.

Figure 3 illustrates the joint training strategy, it omits the skip connection and multi-class-grained scanning in SForest for easy understanding. The former i layer of the cascade forest is adopted for updating the generated sample label, which means that only real dataset join the training process. Then, the problem is how to determine the layer i? Inspired by the cascade forest to determine the number of layers, if the probability $p_k(x)$ of generated samples at the k-th layer has larger loss than the probability $p_{k-1}(x)$ of generated samples at the $(k-1)$-th layer, it proves that the correct classification for generated dataset is not achieved, we need to expand the layers until the following constraint is met.

$$L\left(P_k(x), P_{\text{vizt}}(x)\right) - L\left(P_{k-1}(x), P_{\text{bit}}(x)\right) < \varepsilon$$
$$L\left(P_k(x), P_{k-1}(x)\right) - L\left(P_{k-1}(x), P_{k-2}(x)\right) < \varepsilon \tag{3}$$

In formula 3, x is presents generated samples. The number of the layer i can be determined by formula 3. At the same time the generated samples will be processed from layer 1 to layer i to update the initial label and then make a reliable label. Starting from the $i+1$ layer, the real data and generated data are selected to jointly train the classifier, and the loss calculation formula is the sum of the two parts:

$$loss = (1 - \lambda)loss_{\text{real}} + \lambda loss_{\text{generated}} \tag{4}$$

For a real training data, $\lambda = 0$. For a generated training data, $\lambda = 1$. So our model actually has two types of losses, one for real data and another for generated data.

Using this method, we can deal with more training samples that are located near the real training samples, so the generated data with different styles to regularize the model. By updating the weak label of generated image samples through the former i layer of cascade classifier, which can make generated samples more reliable. Then the classifier will learn exact discriminative features because of more diverse data sets by adding generated samples and encourage the model to be less prone to over-fitting.

4   Experiments

In this section, some small-scale datasets were employed to investigate the effectiveness of the proposed WGSForest model. The goal is to validate that the generated samples in WGSForest can increase the diversity of the original training set.

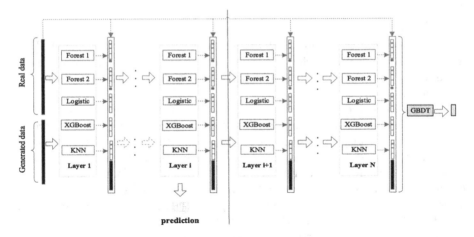

Fig. 3. The joint training strategy of the proposed model. There are two stages: real data train the former i-layer to make the label of generated data more confidence, real data and generated data jointly train the following layer of cascade forest. The dashed line indicates that dataset does not participate in the training, and the solid line represents that dataset participate in the training.

4.1 Configuration

1) InfoGAN model for generating data: We mainly use the infoGAN model and follow the same configuration setting in [11] for fair experimental comparisons. Moreover, we consider the data label in training the discrimination network to improve the performance of infoGAN. In details, we input a fixed dimension random vector including three parts: random noise, a category code $c_1 \sim Cat(K, p = 1/k)$ and three continuous codes $c_2, c_3, c_4 \sim Unif(-1, 1)$. The outputted images are resized to the shape as the original training image and then used in the SForest training.

2) The proposed WGSForest model: The default configuration of SForest is using the same structure as [13]: each level consists of 2 completely-random forests, 2 random forests, 2 logistic regressions, 2 K-neighbours (KNN), 2 Xgboost. Each forest contains 500 trees and neighbours in KNN is 5. Moreover, the top-5 most important features information is transmitted from the trained forest to the next layer. For multi-class-grained scanning, three window sizes are utilized. We use feature window sizes of $[d/16], [d/8], [d/4]$ for d raw feature; the number of the forest is $4 * k$ (k is the class of dataset), each forest containing 1000 trees. In addition, we set $\lambda = 0.2$ and $\varepsilon = 10^{-5}$.

Deep forest uses the default configurations as [14], and we use LSRO [11] and MpRL [12] configuration with the ResNet-50 baseline. For all experiments, we select different sub-datasets by using hierarchical sampling method and compared the classification performance of our model with deep forest, ResNet, LSRO, MpRL and SForest.

4.2 Evaluation

We will validate the performance on grey-scale and color image datasets respectively. Grey-scale image has only two-dimensional features and its form is simple. Color images have three-dimensional characteristics, which are more complex than grey-scale images. In the process of experiment, color images will be significantly different from gray images, which may affect the performance of the experiment. To investigate the effectiveness of SForest, we conducted experiments on some small-scale datasets. In all experiments of this study, we used 5-fold cross-validation to evaluate the overall accuracy of different methods.

MNIST Dataset (Grey-Scale Image): First, we experiment on MNIST dataset, which contains 60000 images of size 28 by 28 for training and 10000 images for testing. Here, we choose 2000/1000, 1000/500 and 500/250 (train/test) sub-datasets for validating the availability of proposed model respectively. This experiment set the extra data ratio with 100%, 100% and 100%, so the generated sample size in this experiment is 2000 (each class own 200 images), 1000 (each class own 100 images), 500 (each class own 50 images) respectively.

As illustration in Table 1, for grey-scale MNIST dataset, the MpRL and the proposed method achieve the highest performance on 2000 training data. However, as the size of the dataset continues to decrease, our method gradually shows its superiority. We observed improvements of +0.68%, 1.17% and +1.57% in 2000/500 1000/500 and 500/250 sub-dataset than SForest, respectively.

Table 1. Validating WGSForest model performance on grey-scale MNIST dataset.

Data	Algorithm					
	Deep forest	ResNet-18	LSRO	MpRL	SForest	WGSForest
2000/1000	98.23%	98.40%	99.38%	**99.50%**	98.82%	**99.50%**
1000/500	96.67%	96.39%	98.80%	98.85%	97.93%	**99.10%**
500/250	96.40%	95.60%	98.80%	98.40%	97.32%	**98.89%**

ORL Dataset (Grey-Scale Image): The ORL dataset contains 400 grey-scale facial images taken from 40 persons, each person has 10 images. We compare it with a CNN consisting of 2 conv-layer with 32 feature maps of 3×3 kernel, and each conv-layer has a 2×2 max-pooling layer followed. We randomly choose 5/7/9 images per person for training and report the test performance on the remaining images. We trained the ResNet18 and ResNet34 on ORL dataset. We generate 1000 samples for each sub-dataset to augment original data.

It can be seen from Table 2 that when the amount of training data is relatively small, the proposed algorithm has a better effect than LSRO and MpRL. When the training data is 360 (9 images), the MpRL effect is slightly better than the proposed algorithm. Possibly because MpRL is in the right place for the dataset

Table 2. Validating WGSForest model performance on grey-scale ORL dataset.

Algorithm	Data		
	5 images	7 images	9 images
Deep forest	91.00%	96.67%	97.50%
CNN	86.50%	91.67%	95.00%
ResNet18	90.00%	93.33%	97.50%
ResNet34	87.50%	95.00%	95.00%
SForest	91.68%	96.81%	97.83%
LSRO	91.50%	95.00%	97.90%
MpRL	93.50%	96.67%	**98.98%**
WGSForest	**94.74%**	**97.82%**	98.58%

parameters. Interestingly, the effect of the original deep forest have almost same performance as LSRO on three sub-datasets.

Cifar10 Dataset (Color Image): The cifar10 dataset contains 50,000 coloured 32 by 32 images of 10 classes for training and 10,000 images for testing. For validating the proposed model in colour image, we select 500/200, 1000/500 and 2000/1000 (train/test) sub-dataset to train the classification model. The generated number of three sub-dataset are 500, 1000 and 2000 (extra data rate is 100%).

As illustration in Table 3, for the colour cifar10 dataset, the proposed method is at a disadvantage compared with the MpRL method, but it has almost accuracy compared with LSRO. Through the results of this experiment, we can see that the processing ability of the deep forest method on the colour dataset is still not good enough. Due to limitation of computational resource, we have not tried a larger model with more grains, forests and trees.

Table 3. Validating WGSForest model performance on colour image cifar10 dataset.

Algorithm	Data		
	500/250	1000/500	2000/1000
Deep forest	36.80%	42.80%	43.90%
ResNet18	44.00%	52.00%	58.70%
ResNet34	40.00%	50.99%	58.30%
SForest	43.80%	50.80%	57.80%
LSRO	49.20%	56.50%	61.70%
MpRL	49.50%	**57.50%**	**63.20%**
WGSForest	**50.04%**	56.00%	61.70%

InfoGAN Images vs. Real Images: To further evaluate the proposed method, we replace the infoGAN images with the real data from MNIST which are viewed as weakly labelled data in training. Since MNIST only contains 1000 images for training, we randomly select 1000 images from the remaining images for the fair comparison. Experimental results are shown in Table 4. We compare the results obtained using 1000 real images from MNIST with the 1000 generated images from infoGAN. We can find the model trained with infoGAN-generated data that assist in the regularization and improve the performance. But the model trained with the real data is slightly better. We can see that the images generated by infoGAN can capture the target distribution. We also observe improvements of +0.67% and 7.47% in MNIST and cifar10 respectively, on the improved deep forest baseline.

Table 4. We add 1000 real images as weakly-labeled images to dataset.

Supplement data	MNIST (1000/500)	Cifar10 (1000/500)
0 (the improved deep forest)	97.83%	50.80%
Real-1000/500	**98.62%**	**57.00%**
InfoGAN-1000/500	98.50%	56.27%

The Impact of Different Extra Data Rate u**:** An interesting issue arises. That is, whether some appropriate values of generated data rate u exist, which enables the classification ability of the proposed model to be better. To explore this issue, further experiments are performed on the dataset (MNIST, ORL, cifar10). MNIST chooses the sub-dataset with 1000/500 (train/test) data, ORL chooses the 7 per person for training, and report the test performance on the remaining images, Cifar10 choose the sub-dataset with 1000/500 (train/test) data. The results are depicted in Table 5, which reveals that for any dataset, the classification performance of the proposed model can be better than SForest, given that u is set to an appropriate value. Our proposed method improved the accuracy of MNIST, ORL and cifar10 by +1.66%, 1.47% and 7.78% respectively when adding 300%, 200% and 200% generated images. But when the value of u becomes bigger, the proposed model not as safe since there are cases where it significantly deteriorates the performance of proposed model. The reason might be that when too many extra training data are generated, the chances of overfitting is enlarged. These results indicate that the weakly labeled images generated by the infoGAN effectively yield improvements over the deep forest model using the joint training strategy. Obviously, the time increases rapidly when extra data sets are added at the beginning, and then slowly increases.

Table 5. Different supplement rate of generated data on three dataset.

Supplement	Dataset					
	MNIST		Orl		cifar10	
	Accuracy	Training time	Accuracy	Training time	Accuracy	Training time
0	97.83%	453 s	96.81%	601 s	50.80%	1201 s
100%	99.10%	840 s	97.82%	1153 s	56.40%	1472 s
200%	99.38%	1072 s	**98.28%**	1471 s	**58.58%**	1695 s
300%	**99.49%**	1312 s	97.80%	1862 s	57.80%	1904 s
400%	99.22%	1484 s	96.87%	2181 s	56.60%	2130 s
500%	99.10%	1543 s	99.22%	2301 s	50.48%	2315 s

5 Conclusion

In this paper, we proposed a new joint training method WGSForest for the generated data by infoGAN. The generated data with weak label were come from infoGAN and were processed by cascade forest to assign a confident label. The real data and the generated data were integrated to jointly train the following cascade forest layer. The experimental results indicated that the performance of the classifier can be improved to some extent by adding the specific proportion of the generated weak label samples, because the infoGAN trained by the real samples can obtain more diverse generated data sets under the real data distribution. Although the proposed model is less effect than the model based on deep learning, this is an attempt to use the deep forest model for data augmentation. In the future, we will continue to research on whether GAN can generate sequence data which is an advantage object that can be processed by the deep forest.

References

1. Hu, G., Peng, X., Yang, Y., Hospedales, T.M., Verbeek, J.: Frankenstein: learning deep face representations using small data. IEEE Trans. Image Process. **27**(1), 293–303 (2017)
2. Chen, M., Shi, X., Zhang, Y., Wu, D., Guizani, M.: Deep features learning for medical image analysis with convolutional autoencoder neural network. IEEE Trans. Big Data (2017)
3. Zagoruyko, S., Komodakis, N.: Learning to compare image patches via convolutional neural networks. In: Proceedings of the IEEE Conference on Computer Vision and Pattern Recognition, pp. 4353–4361 (2015)
4. Goodfellow, I., et al.: Generative adversarial nets. In: Advances in Neural Information Processing Systems, pp. 2672–2680 (2014)
5. Arjovsky, M., Chintala, S., Bottou, L.: Wasserstein GAN. arXiv preprint arXiv:1701.07875 (2017)
6. Radford, A., Metz, L., Chintala, S.: Unsupervised representation learning with deep convolutional generative adversarial networks. arXiv preprint arXiv:1511.06434 (2015)

7. Karras, T., Laine, S., Aila, T.: A style-based generator architecture for generative adversarial networks. In: Proceedings of the IEEE Conference on Computer Vision and Pattern Recognition, pp. 4401–4410 (2019)

8. Chen, X., Duan, Y., Houthooft, R., Schulman, J., Sutskever, I., Abbeel, P.: Info-GAN: interpretable representation learning by information maximizing generative adversarial nets. In: Advances in Neural Information Processing Systems, pp. 2172–2180 (2016)

9. Odena, A.: Semi-supervised learning with generative adversarial networks. arXiv preprint arXiv:1606.01583 (2016)

10. Salimans, T., Goodfellow, I., Zaremba, W., Cheung, V., Radford, A., Chen, X.: Improved techniques for training GANs. In: Advances in Neural Information Processing Systems, pp. 2234–2242 (2016)

11. Zheng, Z., Zheng, L., Yang, Y.: Unlabeled samples generated by GAN improve the person re-identification baseline in vitro. In: Proceedings of the IEEE International Conference on Computer Vision, pp. 3754–3762 (2017)

12. Huang, Y., Xu, J., Wu, Q., Zheng, Z., Zhang, Z., Zhang, J.: Multi-pseudo regularized label for generated data in person re-identification. IEEE Trans. Image Process. **28**(3), 1391–1403 (2018)

13. Zhang, M., Zhang, Z.: Small-scale data classification based on deep forest. In: Douligeris, C., Karagiannis, D., Apostolou, D. (eds.) KSEM 2019. LNCS (LNAI), vol. 11775, pp. 428–439. Springer, Cham (2019). https://doi.org/10.1007/978-3-030-29551-6_38

14. Zhou, Z.H., Feng, J.: Deep forest. arXiv preprint arXiv:1702.08835 (2017)

15. Pathak, D., Krahenbuhl, P., Donahue, J., Darrell, T., Efros, A.A.: Context encoders: feature learning by inpainting. In: Proceedings of the IEEE Conference on Computer Vision and Pattern Recognition, pp. 2536–2544 (2016)

16. Reed, S., Akata, Z., Yan, X., Logeswaran, L., Schiele, B., Lee, H.: Generative adversarial text to image synthesis. arXiv preprint arXiv:1605.05396 (2016)

17. Wu, J., Zhang, C., Xue, T., Freeman, B., Tenenbaum, J.: Learning a probabilistic latent space of object shapes via 3D generative-adversarial modeling. In: Advances in Neural Information Processing Systems, pp. 82–90 (2016)

18. Wen, H., Zhang, J., Lin, Q., Yang, K., Huang, P.: Multi-level deep cascade trees for conversion rate prediction in recommendation system. In: Proceedings of the AAAI Conference on Artificial Intelligence, vol. 33, pp. 338–345 (2019)

19. Wang, H., Tang, Y., Jia, Z., Ye, F.: Dense adaptive cascade forest: a self-adaptive deep ensemble for classification problems. Soft Comput. **24**(4), 2955–2968 (2019). https://doi.org/10.1007/s00500-019-04073-5

20. Pang, M., Ting, K.M., Zhao, P., Zhou, Z.H.: Improving deep forest by confidence screening. In: 2018 IEEE International Conference on Data Mining (ICDM), pp. 1194–1199. IEEE (2018)

21. Guo, Y., Liu, S., Li, Z., Shang, X.: BCDForest: a boosting cascade deep forest model towards the classification of cancer subtypes based on gene expression data. BMC Bioinform. **19**(5), 118 (2018)

An Answer Sorting Method Combining Multiple Neural Networks and Attentional Mechanisms

Liguo Duan, Jin Zhang$^{(\boxtimes)}$, Long Wang, Jianying Gao, and Aiping Li

College of Information and Computer, Taiyuan University of Technology,
Taiyuan 030024, Shanxi, China

Abstract. A deep learning model combining multiple neural networks and attentional mechanisms is proposed to solve the problem of answer ranking. The word vectors of the questions and candidate answers were sent to the Convolutional Neural Network for learning, which is used the Leaky Relu activation function, and the learning results were pieced together with four Attention items, and features in relation to the vocabulary and topic, and then input into the Bidirectional Gated Recurrent Units. After the output results were processed by Multi-layer Perception, the softmax classifier produced the final ranking results. Experimental results indicate satisfactory performance of the model on WikiQACorpus data set with an accuracy of 80.86%.

Keywords: Answer sorting · Attention mechanisms · Attention items · Bidirectional gated recurrent units

1 Introduction

With the rapid development of Internet technology, data information increases rapidly. The public domain question-answer system has become the preferred way for more and more people to obtain information and share knowledge. The community question-answer system, represented by Baidu Zhidao, is widely used in people's life. However, there are many different answers to the same question, and the quality of the answers varies. Therefore, the answer sorting algorithm, which aims at analyzing and filtering the user's question information and returning the accurate answer, plays an important role [26].

As an important part of question answering system, ranking strategy is the key to ensure the quality of question answering system. The answer sorting task is to rank the answers related to the question according to the degree of similarity from the highest to the lowest, in order to choose the best answer [5].

Through the analysis of many question-and-answer information data, it is found that for question-and-answer pairs of texts with different lengths, it is unlikely to extract more information from short texts; while the redundancy of

---

Supported by organization x.

H. Mei et al. (Eds.): BigData 2020, CCIS 1320, pp. 96–111, 2021.
https://doi.org/10.1007/978-981-16-0705-9_7

information in long texts usually makes it difficult to achieve high accuracy due to the varying sorting effects. Therefore, this paper proposes an answer ordering model that combines multiple neural networks and attention mechanism, with the former aiming at learning text features and the latter selecting useful text information to improve accuracy.

The main contributions of this paper are: 1) four commonly used attention models, namely concat, dot-product, scaled dot-product and bilinear, are introduced into the alignment relationship between questions and candidate answers, so as to obtain the Attention terms related to the current candidate answers. 2) a multi-layer neural network model, composed of convolutional neural network, bidirectional gated recurrent units and multilayer perceptron, is constructed with the joint layer incorporating Attention items, characteristics of vocabulary and topics, and other information. 3) CNN (convolutional neural network) is improved and adopted, Pooling operation was performed with 2-max Pooling, and Leaky Relu activation function was introduced to solve the gradient disappearance problem and accelerate the convergence speed. 4) the above method performs well on WikiQACorpus data set.

2   Related Work

At present, the answer-sorting methods can be divided into two categories: one is the traditional machine learning method based on characteristics of vocabulary, syntax and grammar, etc.; and the other one is the new deep learning method based on neural network model. In recent years, some researchers have tried to apply deep learning model to the task of sorting answers and achieved good results. Nie L selected answers by building a knowledge base composed of an offline learning component and an online search component [11]. Severyn A used binary convolutional neural network to generate word vectors of questions and answers respectively, and then calculated the similarity value between the question and candidate answers by similar matrix. The similarity value was spliced with the word vector of the question and candidate answers, with the resulted vector, together with some other features, sent to the softmax layer for classification [15]. Nie YP introduced the bidirectional long short term memory network (LSTM) encoder and decoder to effectively solve the lexical inconsistencies between questions and answers in machine translation tasks, and used the step attention mechanism to make questions focus on a certain part of the candidate answers [12]. Xiang et al. proposed a deep neural network structure for the community question-answer data, which adopted the methods of convolutional neural network, attention-based long short term memory network and conditional random field [19]. Fan et al. used multidimensional feature combination and similarity ranking to improve the community question-answer ranking method of Internet forums, made full use of the information in questions and answers to determine their similarity, and used text-based feature information to determine whether the answers were reasonable [4]. Based on the above research, in order to improve the performance of answer ranking, this paper proposes a new deep learning model that integrates multiple neural networks and attention mechanism.

3   Deep Learning Model Integrating Multiple Neural Networks and Attentional Mechanisms

In the process of sorting the answers, it is not comprehensive to only carry out deep learning of questions and candidate answers, and its unique lexical features, thematic features and other characteristic information also affect the result of answer ranking to a certain extent. Also, it is not necessary to conduct a comprehensive analysis of the candidate answers. Instead, there is a high, probability that some words will determine whether a candidate answer is correct. Therefore, to a certain extent, the introduction of the Attention mechanism affects the answer sorting effect. This paper combines a variety of neural networks and attention mechanisms. To be specific, the convolution neural network was used to extract features of questions and candidate answers, the bidirectional gated recurrent units were used to train the vectors that integrated various information such as Attention items, lexical features and theme features, and the softmax classifier was used to produce the final sorting results following the processing of multi-layer perception. Figure 1 below shows the basic structure of the model.

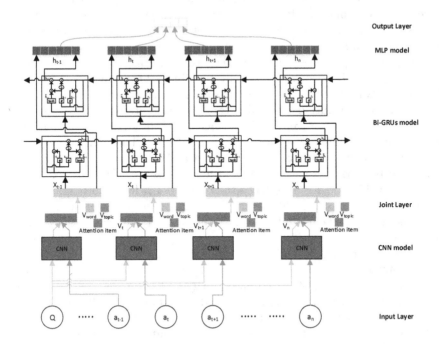

Fig. 1. Network model structure.

3.1 Input Layer

In the input layer, both the questions and candidate answers in the data set are expressed by word vector. The set of word vectors comes from Google news and contains about 3 million English words, with each word having a dimension of 300. Word matching is used for the comparison between each word in the question and candidate answers and the corresponding word in the word vector set, and for the subsequent word replacement. Words that cannot be found are represented by the 0 vector with a dimension 300.

In addition, an analysis of the data in the WikiQA English data set shows that the maxima lengths of the question sentence and the candidate answer are 22 and 98 respectively. Therefore, in order to ensure the robustness of the model, the maximal length of 120 was adopted in the experiment, and 0 was used to complete sentences shorter than 120.

3.2 Convolutional Neural Network

In recent years, as one of the representative algorithms of deep learning, convolutional neural network [2,14] has witnessed a rapid development and been widely applied in computer vision, natural language processing and other fields. The difference between the convolutional neural network and the ordinary neural network is that the convolutional neural network contains a feature extractor, which is composed of convolutional layer and pooling layer [23]. Besides, its convolution and pooling can greatly simplify the model complexity by reducing the parameters of the model. Figure 2 shows an example of the application of convolutional neural network in natural language processing.

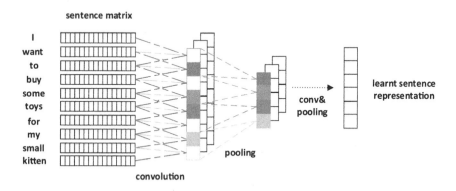

Fig. 2. Application example of convolutional neural network in natural language processing

In this paper, the convolution layer processes the matrix of 120 rows and 300 columns obtained in the input layer, where 120 represents the length of the sentence and 300 represents the dimension of the vector. When the filter is applied to a sentence with a sliding window step size of 2, you get 119 outputs. The operation is shown in Eq. (1).

$$c_i = f\left(\mathbf{W} * v_{i,i+1} + b\right) \qquad (1)$$

Where $\mathbf{W}$ is the convolution layer weight matrix, $v_{i,i+1}$ represents the matrix composed of the word vector from the i-th word to the i + 1-th word when the sliding window slides, b is the bias term, f represents the activation function. The variation leaky-relu [9] based on the Relu function was selected in this paper. Compared with saturated activation functions such as sigmoid and tanh, Relu and its variants, as unsaturated activation functions, can solve the problem of "gradient disappearance" and accelerate the convergence rate. The leaky-relu function was able to handle the Relu function with a slope of 0 when z was less than 0. Its function equation is shown in Eq. (2).

$$f\left(z\right) = \begin{cases} 0.05z, z < 0 \\ z, z \geq 0 \end{cases} \qquad (2)$$

Where 0.05 is the slope, as shown in Fig. 3.

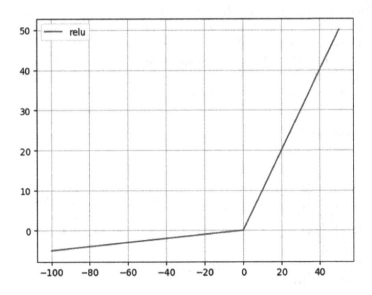

Fig. 3. Leaky-Relu function diagram when k = 0.05

Pooling layer performs pooling operation on the resulted vector $\mathbf{c} = [c_1, c_2, ..., c_{119}]$ after the convolution layer, and the variant 2-max Pooling of maximum Pooling is used. Compared with Max Pooling, 2-max Pooling enables the feature values to be taken from those scoring at top-2, and the original sequence of these feature values is retained. Through pooling operation, the important part of the feature set extracted by each filter can be obtained, which is the important semantic part of the sentence. Figure 4 shows an example of 2-max Pooling.

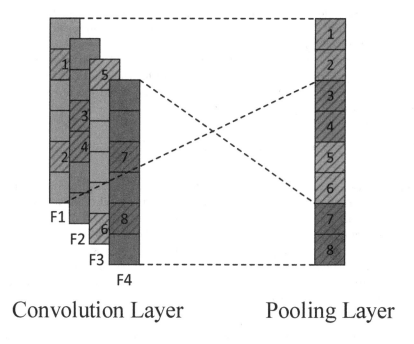

Convolution Layer   Pooling Layer

Fig. 4. 2-max Pooling figure

In addition, in order to prevent overfitting, Dropout mechanism is introduced [13], which can stop the activation values of a number of neurons with a certain probability p when the data are propagated forward, and reactivate them when the data are propagated back.

3.3   Joint Layer

The joint layer combines the vectors learned by the convolutional neural network with the lexical features, topic features and Attention items. The lexical features are obtained by the Stanford CoreNLP toolkit provided by Stanford University, it mainly includes parts of speech and named entity recognition, and there are 36 kinds of part of speech tagging in English. And the topic features are provided by the Title item in the dataset. Both the lexical features and thematic features are encoded by one-hot code.

The Attention term is generated by introducing the Attention mechanism [18,22] on the basis of combining alignment matrix with the bidirectional gated recurrent units model of question and candidate answer. Alignment matrix is based on the fact that when the question-answer system selects candidate answers, it will consider that there is a corresponding relationship between the question and the candidate answers. For example, the common question of "how many" usually corresponds to the cardinality words in the candidate answers, and uses the IBMModel alignment model to find the alignment relationship between the question and the candidate answers. Each word in the question corresponds to all the words in each candidate answer to obtain its alignment relation, and an alignment matrix is generated for each question-answer pair. Let the alignment matrix of the i-th question and answer pair be expressed as $A_i \in \mathbb{R}^{|q_i| \times |a_i|}$, matrix is $|q_i| \times |a_i|$ dimension, $|q_i|$ for the length of the question, and $|a_i|$ for the length of the candidate answer. The definition of each element in A_i is shown in Eq. (3).

$$A_i[t,j] = \begin{cases} decpro[t], align(q_{i,t}, a_{i,j}) \\ 0, other \end{cases} \tag{3}$$

The equation says, if the t-th word in the i-th question is aligned with the j-th word in the candidate answer, then, the alignment probability of the two words, $decpro[t]$, is taken as the value of the corresponding position of the matrix. If there is no alignment, it is filled with zero value. According to the definition of the above equation, the alignment matrix between the question and the candidate answer can be obtained, which indicates the probability that the words in the candidate answers are selected. Table 1 shows an example of the alignment relationship between question and candidate answers.

Table 1. Examples of alignment relationships between questions and candidate answers.

Question	How many stripe flag
Question_pos	WRB JJ NN NN
Answer	50 star flag represent 50 state united state america 13 stripe represent thirteen british colony declare independence kingdom great britain become first state union
Answer_pos	CD NNS NN VBP CD NNS JJ NNS NN CD NNS VBP CD JJ NNS VBD NN NN JJ NN VBD JJ NNS NN
Result	many→CD(0.1971), stripe→VBP(0.1344), flag→NNS(0.1354)

In this paper, the alignment relationship is obtained by using the question and the part of speech of the candidate answer. It can be found that the word many is exactly aligned with the cardinal number word (CD), indicating that the correct answer to the question should be a specific number, and then the corresponding relationship between the question and the answer can be obtained. According to the corresponding relation, the alignment matrix A_i of the question and the candidate answer is generated. Then, candidate answer a_i is sent to the bidirectional gated recurrent units to obtain the hidden state output matrix of candidate answer a_i, and its output matrix is set as $H_{a_i} \in \mathbb{R}^{|a_i| \times 300}$. 300

represents the dimension formed by bidirectional stitching of the dimension of the hidden layer state of the gated recurrent unit. Finally, four common attention models, namely concat model, dot-product model, scaled dot-product model and bilinear model, are introduced into the alignment matrix A_i and hidden state output matrix of candidate answer a_i. The four common attention models [16] are shown in Eqs. (4), (5), (6) and (7).

concat model

$$s\left(A_i, H_{a_i}\right) = \mathbf{v}^T \tanh\left(\mathbf{W} A_i + \mathbf{U} H_{a_i}\right) \tag{4}$$

dot-product model

$$s\left(A_i, H_{a_i}\right) = A_i^T H_{a_i} \tag{5}$$

scaled dot-product model

$$s\left(A_i, H_{a_i}\right) = \frac{A_i^T H_{a_i}}{\sqrt{300}} \tag{6}$$

bilinear model

$$s\left(A_i, H_{a_i}\right) = A_i^T \mathbf{W} H_{a_i} \tag{7}$$

Where $\mathbf{W}$, $\mathbf{U}$ and $\mathbf{v}$ are learnable network parameters. After the calculation, the Attention terms $A_{a_i} \in \mathbb{R}^{|q_i| \times 300}$ related to the current candidate answers were obtained. The calculation process is shown in Fig. 5.

3.4 Bidirectional Gated Recurrent Units

The gated recurrent unit is a variant mode based on the long short term memory network model [7]. Compared with the traditional memory block composed of the forget gate unit, the input gate unit, the output gate unit and the memory cell unit, the gated recurrent unit has only one update gate unit and one reset gate unit. It cancels the memory cell unit that carries out linear self-update, which instead, is carried out directly in the hidden unit by using the gate mechanism [24]. In addition, generally speaking, the gated loop unit is unidirectional, which can only predict the output of the next moment based on the timing information of the previous moment. But in some cases, the output at the current moment is related not only to the previous state, but also to the future state; therefore, a bidirectional gated recurrent unit [3, 21] is proposed, which can process sequences in two directions at the same time. Each node generates two independent output vectors of the gated recurrent unit, and then the two output vectors are spliced as the output of the current node, thus obtaining the information of the current node. In simple terms, it may be regarded as two layers of neural network, with the first layer being the starting input to the sequence from the left, and the second layer being the starting input to the sequence from the right. In terms of text processing, the first layer can be understood as starting from the beginning of the sentence, and the second layer as the input from the last word of the

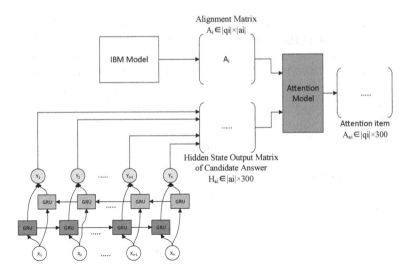

Fig. 5. Calculation process of Attention item

sentence, doing the same thing in reverse as in the first layer. Finally, the two results are spliced. Figure 6 shows the schematic diagram of the Bidirectional Gated Recurrent Units.

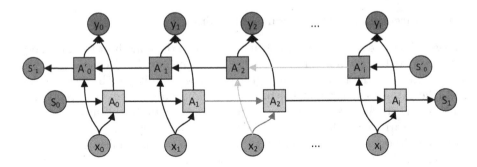

Fig. 6. Bidirectional Gated Recurrent Units

Taking the forward directional gated recurrent unit as an example, the update gate unit is mainly used to control the amount of information of the previous state that is brought into the current state. The state information of the previous moment and the current moment is transformed by linear transformation, and the data obtained after addition is sent to the update gate unit. The larger the value z_t of the update gate unit is, the more state information of the previous moment is brought in, as shown in Eq. (8).

$$z_t = \sigma \left(\mathbf{W_z} \bullet \left[\overrightarrow{h_{t-1}}, x_t \right] \right) \qquad (8)$$

The reset gate unit is used to control the amount of information of the previous state that is written to the current candidate set $\widetilde{h}_t$. Similar to the update gate unit, the lower the value r_t of the reset gate unit is, the less state information is written about the previous moment. The data processing of the reset gate unit is shown in Eq. (9).

$$r_t = \sigma \left(\mathbf{W_r} \bullet \left[\overrightarrow{h_{t-1}}, x_t \right] \right) \tag{9}$$

Since the gated recurrent unit no longer uses a single memory cell unit to store memory information, it directly records the historical state with the hidden cell. The reset gate unit is used to control the data volume of current information and memorized information, and to generate new memory information to continue transmission. Since the output of the reset gate unit ranges within the interval $[0, 1]$, when the reset gate unit is used to control the amount of data that can be transmitted forward by the memory information, the output value of the reset gate unit, 0, means that all the memory information is eliminated, and the value of 1 means that all the memory information can pass through. This can be shown in Eq. (10).

$$\widetilde{h}_t = \tanh \left(\mathbf{W} \bullet \left[r_t \circ \overrightarrow{h_{t-1}}, x_t \right] \right) \tag{10}$$

The hidden state output information $\overrightarrow{h_t}$ is determined by the hidden state information $\overrightarrow{h_{t-1}}$ at the previous moment and the hidden state output $\widetilde{h}_t$ at the current moment, and the amount of data transmitted to the next moment is controlled by the update gate unit. This is shown in Eq. (11).

$$\overrightarrow{h_t} = (1 - z_t) \circ \overrightarrow{h_{t-1}} + z_t \circ \widetilde{h}_t \tag{11}$$

Similarly, the output results of the reverse gated recurrent unit are obtained, and the forward and reverse outputs are spliced, as shown in Eq. (12).

$$\mathbf{h_t} = \left[\overrightarrow{h_t}, \overleftarrow{h_t} \right] \tag{12}$$

3.5   Multi-layer Perceptron

Perceptron network [17] is a feedforward artificial neural network model. It adopts a unidirectional multi-layer structure, in which each layer contains a number of neurons, and there is no mutual connection between the neurons of the same layer. This means the inter-layer information transmission only goes in one direction. Each neuron starts from the input layer, receives the input of the previous layer, and outputs the result to the next layer until the output layer [6,8]. The first layer is called the input layer, the last layer is the output layer, and the middle layer is the hidden layer. The hidden layer can be any layer. Figure 7 shows a multi-layer perceptron with one hidden layer.

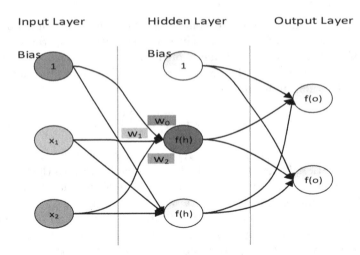

Fig. 7. A Multi-Layer Perception with one hidden layer

During the experiment, $\mathbf{h_t}$ obtained by splicing the bidirectional gated recurrent units layers was sent into the multi-layer perceptron to calculate the predicted score $score_t$ of the current candidate answer a_t, as shown in Eq. (13) [25].

$$score_t = \sigma\left(\mathbf{W}_y\mathbf{h_t} + b_y\right) \tag{13}$$

Where $\mathbf{W}_y$ and b_y represent weight matrix and bias respectively.

3.6 Output Layer

The output layer uses the softmax classifier to process the score corresponding to each candidate answer, and selects the value with the highest probability as the final result. The calculation equation is shown in (14).

$$Softmax_i = e^{score_i} \bigg/ \sum\nolimits_j e^{score_j} \tag{14}$$

4 Experiment and Analysis

4.1 Data Preprocessing

The data set used in this study is from the WikiQA English data set published by Microsoft in 2015. It is a data set for open domain question answering, containing 3,047 questions and 29,258 sentences, with all the questions selected from the search logs of Microsoft's Bing search service to reflect real user needs. The questions searched by each user are connected to a wikipedia page related to the topic of the query question, and each sentence of the abstract is taken as a candidate answer to the question. Manual labeling is used to judge whether all candidate answers are correct or not [20]. Table 2 shows the details of the WikiQA dataset.

Table 2. WikiQA data set information.

	Train set	Dev set	Test set	Total
Question number	2118	296	633	3047
Answer number	20360	2733	6155	29258

4.2 Setting of Hyperparameters

Based on the literature, the filter length in the convolutional neural network layer is set as 3, 4 and 5, the number of filter 50, the dropout ratio 0.5, Adam learning rate 0.01, and batch_size 64. The state of the hidden layer in the bidirectional gated recurrent units is set to 150 dimensions, and the learning rate, dropout rate and the convolutional neural network layer are set to the same extent. RMSprop optimizer is used in the multi-layer perceptron layer, and the dropout ratio is set in accordance with the convolutional neural network layer. The results of the convolution with different filter numbers are shown in Fig. 8.

4.3 Experimental Analysis

Self-comparison experiment is used to analyze the sorting performance of the proposed method. Various different methods are included for comparison, including CNN, CNN+WORD, Bi-GRU, Bi-GRU+WORD+TOPIC, CNN+Bi-GRU, CNN+Bi-GRU+MLP+WORD+TOPIC, and CNN+Bi-GRU+MLP+WORD+TOPIC+ATTENTION.

CNN sents the results of convolution and pooling of the vector representation of the question and the candidate answer to softmax for classification. CNN+WORD, compared with CNN, adds lexical features to the vector representation of questions and candidate answers. Evolving from the gated recurrent unit (GRU), Bi-GRU is a variant of LSTM, which adds a reverse GRU to the forward GRU so that the two independent GRU output vectors are combined as the output of the current node. Bi-GRU+WORD+TOPIC combines Bi-GRU, vector representation of words, lexical features and topic features. CNN+Bi-GRU is a combination of convolutional neural network and bidirectional gated recurrent units, and the output of CNN is taken as the input of Bi-GRU. The method of CNN+Bi-GRU+MLP+WORD+TOPIC introduces multi-layer perceptron to the two-layer deep learning network, and adds various text features like vocabulary, topic and so on. CNN+Bi-GRU+MLP+WORD+TOPIC+ATTENTION is the method proposed in this article. The specific experimental results are shown in Table 3.

On the basis of self-comparison experiment, further comparison experiments are also carried out, including Word Cnt, PV-Cnt, CNN-Cnt, CNNlexical, Bi-LSTM+Lexical+att+Cnt+Topic, and CNN+Bi-GRU+MLP+WORD+TOPIC+ATTENTION.

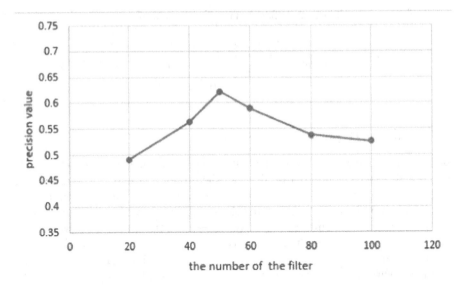

Fig. 8. The results of the different numbers of filters

Table 3. Comparison of experimental results.

Method	Dev set precision	Test set precision
CNN	0.6232	0.6210
CNN+WORD	0.6232	0.6210
Bi-GRU	0.6232	0.6210
Bi-GRU+WORD+TOPIC	0.6232	0.6210
CNN+Bi-GRU	0.6232	0.6210
CNN+Bi-GRU+MLP+WORD+TOPIC	0.6232	0.6210
CNN+Bi-GRU+MLP+WORD+TOPIC+ATTENTION	0.6232	0.6210

Word CNT calculates the number of the same Word in the question and the candidate answer and chooses it as the evaluation criterion. The PV-Cnt method firstly generates vector representations of questions and candidate answers, calculates their semantic similarity, and then combines them with word co-occurrence features. CNN-Cnt combines the results of convolutional neural network and word co-occurrence feature. The method of CNNlexical [10] firstly gets vector representations about questions and candidate answers, then introduces lexical features after the words of the questions and candidate answers are decomposed, and the candidate answers are scored at last. Bi-LSTM+Lexical+att+Cnt judges the candidate answers through adding to the Bi-LSTM model vocabulary characteristics, word co-occurrence characteristics and the dot-product attention model.

The new evaluation index, Mean Average Precision, was used in the comparison. AP is to calculate the average accuracy of judging the correct answer for a problem, while mAP is to find the average for all questions. The calculation method of mAP is shown in Eq. (15).

$$mAP = \frac{1}{|Q_R|} \sum_{q \in Q_R} AP(q) \tag{15}$$

Where AP represents the average accuracy, q represents the total number of samples in a single category, and Q_R represents the number of question categories.

Specific model result pairs are shown in Table 4.

Table 4. Comparison of model results.

Method	WikiQAmAP
Word Cnt	0.4891
PV-Cnt	0.5976
CNN-Cnt	0.6520
CNN$_{lexical}$	0.7058
Bi-LSTM+Lexical+att+Cnt+Topic	0.7891
CNN+Bi-GRU+MLP+WORD+TOPIC+ATTENTION	0.8086

The comparison of the above results clearly shows that the accuracy of the deep learning model proposed in this paper is higher than that of the existing models.

5   Conclusion

This paper proposes a deep learning model that integrates multiple neural networks and attentional mechanisms. The word vectors of the questions and candidate answers were sent to the Convolutional Neural Network for learning, and the learning results were pieced together with four Attention items, the vocabulary features and topic features and input into the Bidirectional Gated Recurrent Units. The output results were processed by Multi-layer Perception, and the final ranking results were obtained by softmax classifier. The model achieves good experimental results on WikiQACorpus data set. In addition, the experimental results show that, apart from the characteristic information of the text, the introduction of the attention mechanism makes the experimental effect significantly better than the simple deep learning model, indicating that the attention mechanism plays a role to some extent.

The next step is to fine-tune the threshold for filtering answers, taking into account the fact that some questions in the dataset do not have right answers. Meanwhile, the latest BERT model [1] will be introduced to further improve the sorting effect.

References

1. Alberti, C., Lee, K., Collins, M.: A BERT baseline for the natural questions (2019)
2. Bai, X., Shi, B., Zhang, C., Cai, X., Qi, L.: Text/non-text image classification in the wild with convolutional neural networks. Pattern Recognit. **66**, 437–446 (2016)
3. Deng, Y., Wang, L., Jia, H., Tong, X., Li, F.: A sequence-to-sequence deep learning architecture based on bidirectional GRU for type recognition and time location of combined power quality disturbance. IEEE Trans. Ind. Inf. **15**(8), 4481–4493 (2019)
4. Fan, H., Ma, Z., Li, H., Wang, D., Liu, J.: Enhanced answer selection in CQA using multi-dimensional features combination. Tsinghua Sci. Technol. **24**, 346–359 (2019)
5. Geerthik, S., Gandhi, K.R., Venkatraman, S.: Respond rank: improving ranking of answers in community question answering. Int. J. Electr. Comput. Eng. **6**(4), 1889–1896 (2016)
6. Goay, C.H., Aziz, A.A., Ahmad, N.S., Goh, P.: Eye diagram contour modeling using multilayer perceptron neural networks with adaptive sampling and feature selection. IEEE Trans. Compon. Packag. Manuf. Technol. **9**, 2427–2441 (2019)
7. Greff, K., Srivastava, R.K., Koutník, J., Steunebrink, B.R., Schmidhuber, J.: LSTM: a search space odyssey. IEEE Trans. Neural Netw. Learn. Syst. **28**(10), 2222–2232 (2016)
8. Li, Y., Yang, H., Lei, B., Liu, J., Wee, C.: Novel effective connectivity inference using ultra-group constrained orthogonal forward regression and elastic multilayer perceptron classifier for mci identification. IEEE Trans. Med. Imaging **38**(5), 1227–1239 (2019)
9. Liu, Y., Wang, X., Wang, L., Liu, D.: A modified leaky Relu scheme (MLRS) for topology optimization with multiple materials. Appl. Math. Comput. **352**, 188–204 (2019). https://doi.org/10.1016/j.amc.2019.01.038, http://www.sciencedirect.com/science/article/pii/S0096300319300475
10. Miao, Y., Yu, L., Blunsom, P.: Neural variational inference for text processing. In: Computer Science, pp. 1791–1799 (2016)
11. Nie, L., Wei, X., Zhang, D., Wang, X., Gao, Z., Yang, Y.: Data-driven answer selection in community GA systems. IEEE Trans. Knowl. Data Eng. **29**(6), 1186–1198 (2017)
12. Nie, Y., Han, Y., Huang, J., Jiao, B., Li, A.: Attention-based encoder-decoder model for answer selection in question answering. Front. Inf. Technol. Electron. Eng. **18**(4), 535–544 (2017). https://doi.org/10.1631/FITEE.1601232
13. Poernomo, A., Kang, D.K.: Biased dropout and crossmap dropout: learning towards effective dropout regularization in convolutional neural network. Neural Netw. **104**, 60–67 (2018). https://doi.org/10.1016/j.neunet.2018.03.016, http://www.sciencedirect.com/science/article/pii/S0893608018301096
14. Qiu, N., Cong, L., Zhou, S., Wang, P.: Barrage text classification with improved active learning and CNN (2019)
15. Severyn, A., Moschitti, A.: Learning to rank short text pairs with convolutional deep neural networks. In: The 38th International ACM SIGIR Conference (2015)
16. Si, Z., Fu, D., Li, J.: U-Net with attention mechanism for retinal vessel segmentation. In: Zhao, Y., Barnes, N., Chen, B., Westermann, R., Kong, X., Lin, C. (eds.) ICIG 2019. LNCS, vol. 11902, pp. 668–677. Springer, Cham (2019). https://doi.org/10.1007/978-3-030-34110-7_56

17. Wang, Y., et al.: A clinical text classification paradigm using weak supervision and deep representation. BMC Med. Inf. Decis. Making **19**(1), 1–13 (2019)
18. Wen, J., Tu, H., Cheng, X., Xie, R., Yin, W.: Joint modeling of users, questions and answers for answer selection in CQA. Expert Syst. Appl. **118**, 563–572 (2018)
19. Xiang, Y., Chen, Q., Wang, X., Qin, Y.: Answer selection in community question answering via attentive neural networks. IEEE Signal Process. Lett. **24**(4), 505–509 (2017)
20. Yang, Y., Yih, S.W., Meek, C.: WikiQA: a challenge dataset for open-domain question answering. In: Proceedings of the 2015 Conference on Empirical Methods in Natural Language Processing (2015)
21. Yu, W., Yi, M., Huang, X., Yi, X., Yuan, Q.: Make it directly: event extraction based on tree-LSTM and Bi-GRU. IEEE Access **8**, 14344–14354 (2020)
22. Yuan, W., Wang, S., Li, X., Unoki, M., Wang, W.: A skip attention mechanism for monaural singing voice separation. IEEE Signal Process. Lett. **26**(10), 1481–1485 (2019)
23. Zeng, D., Dai, Y., Li, F., Wang, J., Sangaiah, A.K.: Aspect based sentiment analysis by a linguistically regularized CNN with gated mechanism. J. Intell. Fuzzy Syst. **36**, 3971–3980 (2019)
24. Zhang, Y., et al.: Chinese medical question answer selection via hybrid models based on CNN and GRU. Multimedia Tools Appl. **79**(21), 14751–14776 (2019). https://doi.org/10.1007/s11042-019-7240-1
25. Zhou, X., Hu, B., Chen, Q., Wang, X.: Recurrent convolutional neural network for answer selection in community question answering. Neurocomputing **274**, 8–18 (2018)
26. Zhu, N., Zhang, Z., Ma, H.: Ranking answers of comparative questions using heterogeneous information organization from social media. Signal Image Video Process. **13**(7), 1267–1274 (2019). https://doi.org/10.1007/s11760-019-01465-w

Optimal Subspace Analysis Based on Information-Entropy Increment

Zhongping Zhang[1,2(✉)], Iiaojiao Liu[1,2] (iD), Yuting Zhang[1,2], Jiyao Zhang[1,2], and Mingru Tian[2,3]

[1] College of Information Science and Engineering, Yanshan University, Qinhuangdao 066004, China
1129124795@qq.com
[2] The Key Laboratory for Computer Virtual Technology and System Integration of Hebei Province, Yanshan University, Qinhuangdao 066004, China
[3] Hebei Education Examinations Authority, Shijiazhuang 050000, China

Abstract. The data structure is becoming more and more complex, and the scale of the data set is getting larger and larger. The strong limitations and instability in the high-dimensional data environment is showed in traditional outlier detection method. To solve the problems, an Optimal subspace Analysis based on Information-entropy Increment is proposed. The concepts such as mutual information and dimensional entropy to re-portrait the indicators that measure the pros and cons of subspace clustering, optimize the objective function of the clustering subspace, and obtain the optimal subspace. According to the idea of dividing the information entropy increment by one, the entropy outlier score is proposed as a metric to detect outliers in the optimal subspace. Finally, experiments verify the effectiveness of the algorithm.

Keywords: Outliers · High-dimensional data · Dimensional entropy · Mutual information · Entropy increment

1 Introduction

As an important branch of data mining, outlier detection has always attracted the attention of researchers. Outlier detection is to analyze the target data set, aiming to obtain records that have abnormal behaviors or contain characteristic information that is quite different from other records, and obtain valuable information by analyzing these outliers. The core idea of outlier detection is to first create a normal pattern in the data, and then assign an outlier score to the degree of deviation from this normal pattern for each point [1]. The focus is to explore the deviation from the normal data pattern in the data and then dig out more valuable information. Outlier detection has been applied in many fields, such as intrusion detection, financial fraud detection, medical and public health detection, weather forecasting, etc.

At present, domestic and foreign scholars have done a lot of exploration on the analysis and mining of outliers. Common outlier detection methods include the density-based outlier detection algorithm LOF [2] proposed by Breunig et al., the distance-based

H. Mei et al. (Eds.): BigData 2020, CCIS 1320, pp. 112–122, 2021.
https://doi.org/10.1007/978-981-16-0705-9_8

algorithm FindAllOutsD [3–5] proposed by Knorr et al., the depth-based algorithm DEEPLOC [6] proposed by Johnson et al.,and the information-based outlier detection algorithm [7–10] proposed by Wu S et al. The above methods have good adaptability to ordinary low-dimensional data, but with the explosive growth of data, the phenomenon of "data disaster maintenance" has become an urgent problem to be solved. Eliminating the influence of irrelevant dimensions on data mining is outlier mining Important premise.

Aiming at the outlier detection of high-dimensional data sets, an optimal subspace analysis based on information-entropy increment is proposed. First, lead into mutual information, dimensional entropy and other definitions to re-portrait an index that measures the pros and cons of subspace clustering, optimize the objective function for measuring clustering subspace, thereby eliminating the influence of irrelevant dimensions on outlier detection, and then obtaining the optimal subspace. Then according to the idea of dividing the information entropy increment, the entropy outlier score is proposed, and the outlier detection is performed in the optimal subspace.

2 Related Definitions

Definition 1: Entropy is a thermodynamic function that expresses the disorder of microscopic particles in the system. The symbol is S, which is expressed as $S = k \log \Omega$ [11]. k is called Boltzmann's constant, and Ω in the logarithmic term is the number of microscopic states of the system.

Definition 2: Information entropy is a measure of the average uncertainty of random variables [12]. Suppose X is a random variable, its value set is S(X), P(X) represents the probability of X possible values, then the information entropy of X is defined as E(X), as shown in formula (1).

$$E(X) = - \sum_{x \in S(X)} P(x) \log(P(x)) \tag{1}$$

Definition 3: Dimensional entropy Suppose the number of data in the data set D is m and the dimension is n. Project D to the dimension i to obtain a set of projection D_i. The projection D_i is equally spaced into b segments, and the length of each segment is $(\max(D_i) - \min(D_i))/b$, count the number of data points m_j $(0 \leq j \leq m)$ in each segment, from which a set of corresponding probabilities $p_j = m_j/m$ can be obtained, and then the dimension entropy of dimension i can be obtained [13], as shown in formula (2).

$$ent_i = - \sum_{j=1}^{m} p_j \log p_j \tag{2}$$

Because of "uniform distribution with maximum entropy", we can get

$$ent_{max} = -b \cdot \left(\frac{1}{b} \cdot \log \left(\frac{1}{b} \right) \right) = - \log \left(\frac{1}{b} \right)$$

Knowing that the size of dimension entropy can indicate the degree of uneven distribution of data. A critical coefficient α can be set, use $\alpha \cdot ent_{max}$ as the critical value of entropy. When enti is less than the critical value, it means that the data of this dimension is not uniformly distributed, that is, there may be points farther from the normal cluster in this dimension. These dimensions are filtered out to construct an optimal subspace for finding outliers.

Definition 4: Mutual information (MI for short) [14] is a measure of the interdependence between variables. The mutual information of two discrete random variables X and Y can be defined as Eq. (3).

$$I(X;Y) = \sum_{y \in Y} \sum_{x \in X} p(x,y) \log(\frac{p(x,y)}{p(x)p(y)})$$
$$= E(x) + E(y) - E(x,y)$$

(3)

The mutual information between all individual dimensions in the data set is defined as formula (4).

$$I(\{x_1, ..., x_n\}) = \sum_{i=1}^{n} E(x_i) - E(x_i, ..., x_n)$$

(4)

We can use mutual information to express the correlation between dimensions, calculate the mutual information between each dimension in the optimal subspace, set a threshold β, filter out the dimensions with mutual information value greater than β, and get the calculation of outliers Optimal subspace.

3   Optimal subspace Analysis Based on Information-Entropy Increment

3.1   Algorithm Ideas and Related Definitions

From the perspective of information entropy describing the degree of uneven distribution of the data space, the more scattered the data distribution, the greater the entropy value. In the data set, it is precisely because of the existence of outliers that the entropy of the data space will be larger in comparison. Therefore, information entropy can be used to describe the uneven distribution of the data space [15]. According to the idea of dividing the information entropy increment, the concept of entropy outlier score is proposed. The entropy outlier score describes the outlier degree of the data in the optimal subspace. The following is the definition of entropy outlier score and its properties.

Definition 5: Entropy outlier score If a data point p is removed from the data set, the amount of change in information entropy before and after the point p is excluded from the data space is called the entropy outlier score of the data point p, which is recorded as SCR(p).

$$SCR(p) = E(X_i) - E(X_i^{-p})$$

(5)

Since information entropy can describe the degree of uneven distribution of the data space, the more scattered the data distribution, the greater the entropy value. Therefore, if the outliers are removed, the degree of uneven data distribution will be weakened and the information entropy will decrease. Therefore, the entropy outlier score has the following properties:

Property 1: When an outlier p is removed, the entropy outlier score is greater than or equal to 0, that is, $SCR(p) \geq 0$.

Property 2: When a normal data point p is removed, the entropy outlier score is less than or equal to 0, that is, $SCR(p) \leq 0$.

3.2 Algorithm Description

Aiming at the "data disaster maintenance" problems in outlier detection in high-dimensional data space. This paper proposes an optimal subspace analysis based on information-entropy increment. In the data processing stage, two-step screening of the subspace is carried out to eliminate the influence of irrelevant dimensions on the detection process. In the outlier detection stage, a new outlier detection index, entropy outlier score, is proposed to describe the degree of data outlier. The following are the core steps of the algorithm.

Step 1: For each dimension in the high-dimensional data set, project the data to the dimension, determine the length of the dimension, and then set the number of cutting segments to cut the projection equidistantly, and count the number of data points on each segment to obtain the corresponding probability value of each segment. Use the entropy solution formula to find the dimensional entropy of each dimension. Set a critical coefficient α, and use $\alpha \cdot ent_{max}$ as the critical value of entropy. When ent_i is less than the critical value, it means that the data of this dimension is not uniformly distributed, that is, there may be points farther from the normal cluster in this dimension. Filter these dimensions Come out, construct a superior subspace for finding outliers.

Step 2: Obtaining the optimal subspace is equivalent to obtaining non-redundant attributes. In addition to the information entropy of a single dimension itself, we also need to consider the correlation between dimensions. Therefore, the concept of mutual information is introduced. The greater the mutual information value, the greater the correlation between each other, the calculation formula of mutual information is used in the optimal subspace to calculate the mutual information value of each dimension, and then the threshold β is set to filter out the dimensions with mutual information value greater than β to further obtain the calculated outlier The optimal subspace of points.

Step 3: Entropy represents the uneven distribution of matter. Therefore, the change in entropy after removing a certain point can be used to describe the degree of outlier. If the removed point is an outlier, the entropy will decrease, otherwise the entropy will increase. Perform outlier detection in the optimal subspace. When a certain point is removed, the amount of change before and after entropy is defined as the outlier score. The outlier score of each data point is calculated and sorted, and the first n data points are output as the outlier group points.

4 Algorithm Process

According to the algorithm description and related definitions, the Optimal subspace Analysis based on Information-entropy Increment (OSEI) algorithm proposed in this paper. The algorithm is as follows. In the experiment, the critical coefficient α and the threshold β are set by pre-analyzing the sampling data of the data set, and the cutting distance b is randomly selected in [10, 15].

algorithm: Optimal   subspace   Analysis   based   on   Information-entropy Increment(OSEI)

Input: data set X, cutting distance b, critical coefficient α, threshold β;

Output: Subspace S, $C_{outliers}$.

OSEI（X,b,α,β）

BEGIN

1) Initialization:$C_{outliers}=\Phi$,$S1=\Phi$, $S2=\Phi$;

2) For each $i\in$（1-n）

3) $ent_i = -\sum_{j=1}^{m} p_j \log p_j$;//According to definition 3, calculate the dimensional entropy of each dimension in the data set X//

4) $\alpha \bullet ent_{max}$;//Then calculate the critical value according to the critical coefficient α//

5) If $ent_i < \alpha \bullet ent_{max}$

6)   put i into S_1 ;//Put the dimension smaller than the critical value into the optimal subspace S_1//

7) End If

8) End For

9) For each $i,j\in S_1$

10) $I(X;Y) = \sum_{y\in Y}\sum_{x\in X} p(x,y)\log(\frac{p(x,y)}{p(x)p(y)})$;//According to the formula in

$= E(x)+E(y)-E(x,y)$

Definition 4, calculate the mutual information value between every two dimensions//

11) If I（i,j）$>\beta$

12)   put i,j into S_2;//Put dimensions greater than the threshold into the optimal subspace S_2//

13) End If

14) End For

15) For each $p\in X$

16) $SCR(p) = E(X_i)-E(X_i^{-p})$ $i\in S2$;//According to definition 5, calculate the entropy outlier score of each data point in the optimal subspace S_2//

17) If SCR(p)>0

18)   put p into $C_{outliers}$;//Put data points with an outlier score greater than 0 into the outlier set//

19) End If

20) End For

21) Output $C_{outliers}$.

END

5 Analysis of Algorithms

Correctness Analysis

In the OSEI algorithm, steps 2)–3) are to calculate the dimensional entropy of each dimension. The formula (2) entropy calculation formula is correct, and the formula (3) dimensional entropy calculation is correct. The dimensional entropy is screened according to the threshold set in the literature [8], so the selection of the optimal subspace is correct and reasonable. Step 9)–13) is to calculate the mutual information value between every two dimensions, The formula (4) calculation formula of mutual information is correct. Mutual information can express the correlation between two dimensions, so the selection of the optimal subspace is correct and reasonable. Step 15)–20) is to calculate the outlier score in the optimal subspace. According to that the entropy can express the uneven distribution of the data, Use formula (2) to calculate the value of entropy. The idea of dividing the entropy increment is correct. The change of entropy before and after removing data points can indicate the degree of uneven distribution of data, so the outlier score is correct.In summary, the OSEI algorithm is correct.

Time Complexity Analysis

The time cost of OSEI algorithm is mainly divided into two parts, the calculation of the optimal subspace and the calculation of the outlier score. Step 2)–13) is the calculation of the optimal subspace, when calculating the dimensional entropy, the data needs to be divided into equal distances to obtain the data distribution probability and then calculate the information entropy. The time complexity of this step is $O(d * n)$, where d is the number of dimensions, n is the number of data points. Mutual information calculation can be completed in only 1 cycle, so the time cost is negligible, so the optimal subspace calculation time complexity is $O(d * n)$. The calculation process of outlier score is steps 15)–20), need to calculate the information entropy value of each removed point and the amount of change before and after in the entire data set. The time complexity is $O(k * n * n)$, and k represents the dimension of the subspace after screening. Therefore, the total time complexity is: $O(d * n) + O(k * n * n) = O(kn^2)$.

6 Experiment Analysis

6.1 Experimental Environment and Data Set Information

The algorithm proposed in this paper is verified on the pycharm platform. The experimental environment configuration of the algorithm is composed of two parts: hardware environment configuration and software environment configuration.

Hardware Environment Configuration: CPU 3.9 GHz, memory capacity is 8.00 GB, hard disk capacity is 1 TB.

Software Environment Configuration: 64-bit Windows 10 operating system. The development tool used is pycharm; the experimental program is written in python language and the version is 3.5.2.

The parameter values of the comparison algorithm and the algorithm in this paper are as follows. The OSEI algorithm parameters take default parameters: critical coefficient $\alpha = 3$, threshold $\beta = 0.1$; LOF algorithm parameters take default parameters: $k = 40$; STCS algorithm parameters take default parameters: entropy filter value $w = 3$, threshold $MI = 0.1, k = 12$.

In this experiment, four UCI real data sets are used, Cancer, WBC, Robotnavigation, and WDBC. The specific attributes of the data sets are shown in Table 1.

Table 1. Data set information

Data set	Data amount	Dimension	Outliers number	Outlier ratio
Cancer	683	9	16	2.34%
WBC	683	9	14	2.05%
Robotnavigation	2983	25	30	1.01%
WDBC	372	30	15	4.03%

6.2 Algorithm Performance Evaluation and Experimental Results

In order to verify the effectiveness of the algorithm, this paper will test and compare with the STCS algorithm [16] and the LOF [17] algorithm under different data sets. STCS algorithm is an outlier detection algorithm based on subspace. First, the redundant attributes are filtered out from the perspective of information entropy, and then mutual information is used to measure the correlation between non-redundant attributes, and attributes with correlation greater than the threshold are selected to determine the strong correlation subspace. For the data objects in the strong correlation subspace, using the SLOF (Simplified Local Outlier Factor) algorithm, the formula for calculating the outlier score is given, and then the N data objects with the largest outlier degree are selected as the outlier data. The LOF algorithm is a classic density-based outlier detection algorithm. It uses the local outlier factor to calculate the degree of outlier. The distance between objects in the data space is calculated through all dimensions, and then Then calculate the reachable density of the object, and finally judge the outlier by the degree of local outlier. For the performance evaluation of the algorithm, this paper uses 4 indicators [18], namely AUC (Area Under Curve), recall rate, accuracy rate and weighted evaluation index value F to evaluate the detection results. AUC is the area under the ROC curve. Suppose PN is the number of true outliers included in the data set X, TP is the number of outliers that are correctly detected by the algorithm, and FP is the number of outliers that are incorrectly detected by the algorithm. The recall rate (Re), accuracy rate (Pr) and weighted evaluation index value F are defined as follows.

$$Re = \frac{TP}{PN} \tag{6}$$

$$Pr = \frac{TP}{TP + FP} \tag{7}$$

$$F = \frac{2Re \times Pr}{Re + Pr} \tag{8}$$

The maximum value of Re, Pr and F is 1, and the minimum value is 0. The larger the value of Re, Pr and F, the better the outlier detection result.

The AUC values of the three algorithms under different data sets are shown in Fig. 1. It can be seen that the AUC value of the OSEI algorithm in this article has reached above 0.8 under the four data sets, and all are higher than the other two algorithms. The STCS algorithm is better than the LOF algorithm under the Robotnavigation and WDBC data sets, but it is slightly worse than the LOF algorithm under the Cancer and WBC data sets. It shows that the STCS algorithm performs better in high-dimensional data sets, and the LOF algorithm performs better in low-dimensional. However, in terms of the overall trend, whether it is a high-dimensional data set or a data set with a slightly lower dimensionality, the OSEI algorithm in this article has the best effect, which also shows that the algorithm in this article has strong adaptability.

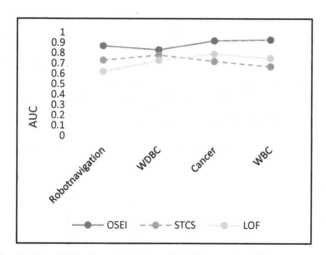

Fig. 1. The AUC values of the three algorithms under different data sets

The recall rate Re of the three algorithms under different data sets is shown in Fig. 2. It can be seen that the OSEI algorithm in this article has a recall rate of 1 under the four data sets, but the recall rates of the other two algorithms fluctuate under different data sets, especially in the Robotnavigation high-dimensional data set, the recall rates of the STCS algorithm and the LOF algorithm are both lower than 0.1, and the effect is relatively poor. The LOF algorithm has a recall rate of 1 under the Cancer and WBC data sets, and the STCS has a recall rate of 0.86 under the WDBC data set. On the whole, the OSEI algorithm recall rate is 1 and very stable, while the STCS algorithm and the LOF algorithm show strong uncertainty.

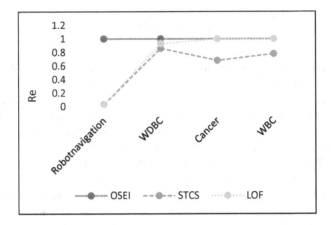

Fig. 2. The recall rate of the three algorithms under different data sets

The accuracy Pr of the three algorithms under different data sets is shown in Fig. 3. It can be seen that the accuracy of the OSEI algorithm in this article is much higher than the STCS algorithm and the LOF algorithm under the four data sets. Especially in the Robotnavigation high-dimensional data set, the accuracy of the STCS algorithm and the LOF algorithm only reached about 0.005, while the OSEI algorithm can reach 0.3, which further shows that the algorithm in this paper is effective and feasible for outlier mining of high-dimensional data sets.

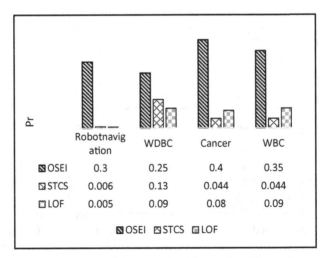

Fig. 3. The accuracy of the three algorithms under different data sets

The F values of the three algorithms under different data sets are shown in Fig. 4. It can be seen that the comprehensive evaluation index F value of the algorithm in this paper is much larger than the STCS algorithm and the LOF algorithm under the four data sets, and it performs more obviously under the Robotnavigation high-dimensional data set. With the change of the dimensions and the number of data, the F value of OSEI algorithm is always higher than that of STCS algorithm and LOF algorithm, so the algorithm in this paper has strong scalability.

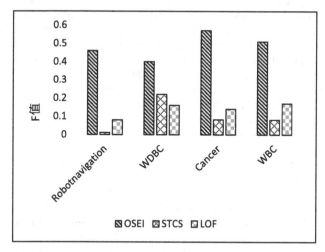

Fig. 4. The F value of the three algorithms under different data sets

7 Conclusion

Outlier mining for high-dimensional data sets is an important field in current data mining research. This paper lead the idea of dividing the increment of information entropy into the outlier detection. The first step of dimensionality reduction is carried out through dimensional entropy to eliminate the influence of irrelevant attributes on outlier detection, use mutual information to further construct the optimal subspace for outlier detection, at the same time, the entropy increment before and after removing each data point is used as a new metric to describe the outlier degree of the data point. Experimental verification shows that the recall rate of the algorithm in this paper can reach 1 under data sets with different dimensions and different numbers of data. The three performance indicators of accuracy, AUC value and F value are also much higher than the comparison algorithm. Therefore, it can be proved that the algorithm in this paper performs well under high-dimensional data sets, has high accuracy and strong scalability, and is effective and feasible for solving the problem of outlier detection in high-dimensional data space.

References

1. Gautam, B., Koushik, G., et al.: Outlier detection using neighborhood rank difference. Pattern Recogn. Lett. **60**, 24–31 (2015)
2. Breunig, M.M., Kriegel, H.P., Ng, R.T., et al.: LOF: identifying density-based local outliers. In: Hen, W.D.C., Naught, J.F., Bernstein, P.A. (eds.) Proceedings of the 2000 ACMSIGMOD International Conference on Management of Data, pp. 93–104. ACM, New York (2000)
3. Kontaki, M., Gounaris, A., Papadopoulos, A.N., et al.: Efficient and flexible algorithms for monitoring distance-based outliers over data streams. Inf. Syst. **55**, 37–53 (2016)
4. Clamond, D., Dutykh, D.: Accurate fast computation of steady two-dimensional surface gravity waves in arbitrary depth. J. Fluid Mech. **844**, 491–518 (2018)
5. Wu, S., Wang, S.R.: Information-theoretic outlier detection for large-scale categorical data. IEEE Trans. Knowl. Data Eng. **25**(3), 589–602 (2013)
6. Chi, Z., Dong, L., Wei, F., et al.: InfoXLM: an information-theoretic framework for cross-lingual language model pre-training. **32**, 154–159 (2020)
7. Coccarelli, D., Greenberg, J.A., Mandava, S., et al.: Creating an experimental testbed for information-theoretic analysis of architectures for x-ray anomaly detection. In: SPIE Defense + Security, pp. 69–72 (2017)
8. Zhang, Z., Qiu, J., Liu, C., et al.: Outlier detection algorithm based on clustering outlier factor and mutual density. Comput. Integr. Manuf. Syst. **2019**(9), 2314–2323
9. Zhang, Z., Fang, C.: Subspace clustering outlier detection algorithm based on cumulative total entropy. Comput. Integr. Manuf. Syst. **21**(8), 2249–2256 (2015)
10. Li, J., Zhang, C., Fan, H.: Swarm intelligent point cloud smoothing and denoising algorithm. Comput. Integr. Manuf. Syst. **17**(5), 935–945 (2011)
11. Department of Inorganic Chemistry, Beijing Normal University, Central China Normal University, Nanjing Normal University. Inorganic Chemistry, pp. 222–227. Higher Education Press, Beijing (2002)
12. Shannon, C.E.: A mathematical theory of communication. ACM SIGMOBILE Mob. Comput. Commun. Rev. **5**(1), 3–55 (2001)
13. Liao, L., Luo, B.: Outlier detection algorithm based on dimensional entropy. Comput. Eng. Des. **40**(4), 983–988 (2019)
14. Zhang, J., Sun, Z., Yang, M.: Mass data incremental outlier mining algorithm based on grid and density. Comput. Res. Dev. **48**(5), 823–830 (2011)
15. Feng, J., Sun, Y.F., Cao, C.: An Information Entropy-Based Approach to Outlier Detection in Rough Sets. Pergamon Press Inc, Oxford (2010)
16. Li, J., Xun, Y.: Strong correlation subspace outlier detection algorithm. Comput. Eng. Des. **38**(10), 2754–2758 (2017)
17. Duan, L., Xiong, D., Lee, J., et al.: A local density based spatial clustering algorithm with noise. In: IEEE International Conference on Systems, pp. 599–603. IEEE (2007)
18. Ning, J., Chen, L., Luo, Z., Zhou, C., Zeng, H.: The evaluation index of outlier detection algorithm. Comput. Appl. **27**(11), 1–8 (2020)

Link Prediction of Attention Flow Network Based on Maximum Entropy Model

Yong Li[1], Jingpeng Wu[1]([⊠]), Zhangyun Gong[1], Qiang Zhang[1], Xiaokang Zhang[2],
Fangqi Cheng[3], Fang Wang[4], and Changqing Wang[5]

[1] College of Computer Science and Engineering, Northwest Normal University, Lanzhou, China
[2] Lanzhou Qidu Data Technology Co., Ltd., Lanzhou, China
[3] School of Computer Science, Beihang University, Beijing, China
[4] 32620 Army, Xining, Qinghai, China
[5] DNSLAB, China Internet Network Information Center, Beijing, China

Abstract. Attention flow network is a new and important branch of network science. Most of the work in this field are devoted to discovering common patterns in the attention flow network and revealing the basic mechanisms and evolution laws of the world wide web. The link prediction algorithm of attention flow network has important theoretical significance and application value but there is little research. Most of the existing link prediction algorithms are based on undirected unweighted networks, which are not suitable for attention flow networks. Because maximum entropy model without extra independent assumptions and constraints, this paper proposes a link prediction algorithm of attention flow network based on the maximum entropy model (MELP). Compared with other algorithms, the experimental results show that: 1) MELP is superior to other algorithms in accuracy, recall rate, F1 value and ROC curve; 2) MELP is more excellent in AUC value. MELP algorithm can provide an entirely new solution to the precision delivery of Internet advertisements, the perception of users' risky behaviors and other applications.

Keywords: Network Science · Attention flow network · Link prediction · Maximum entropy

1 Introduction

In the late 1990s, the discovery of the WS model in the "small world" [1] and BA model in the "scale free" phenomenon [2] inspire a research boom about complex networks and network science [3]. Attention flow network is a kind of directed and weighted network with attributes of nodes. It is an important research branch of network science. It shows the users' clickstream among different websites. The research on link prediction of attention flow network has important theoretical value for the fields of long-range evolution of online users' interest, the temporal and spatial characteristics of human behavior and the evolution dynamics of Websites. It also has application value for commercial fields such as the precision delivery of Internet advertisements and the perception of users' risky behaviors. However, as far as we know, there are few researches in this aspect.

© Springer Nature Singapore Pte Ltd. 2021
H. Mei et al. (Eds.): BigData 2020, CCIS 1320, pp. 123–136, 2021.
https://doi.org/10.1007/978-981-16-0705-9_9

This paper proposes a link prediction algorithm based on the maximum entropy model. Compared with support vector machine, naive Bayes and perceptron algorithm, it has more excellent link prediction performance. It can be applied to the fields of precision delivery of Internet advertisements and the perception of users' risky behaviors, etc. It is also important to guide further study of attention flow network.

2   Related Work

2.1   Link Prediction

The main task of link prediction algorithm is to estimate the possibility [4] of link between two nodes according to the observed link and node attributes. Link prediction algorithm can solve many practical problems in life [5–7].

In [8], Lü and Zhou reviewed these classical link prediction algorithms. These algorithms predict the links through network topology properties. Most of them are based on undirected unweighted graph. They are not suitable for attention flow network.

Social networks such as Twitter and Sina Micro-blog have become increasingly complex in recent years. These classical link prediction algorithms are unable to handle further information. Therefore, more and more scholars begin to solve link prediction by machine learning algorithms. If edges exist in the network, take it as a positive sample. Otherwise, as a negative sample. In this way, link prediction can be transformed into a two-category problem. Then, the problem can be solved by machine learning algorithms [9, 10]. These algorithms usually take a lot of additional information into account besides the network structure characteristics. It leads to the low universality of these algorithms and they are not suitable for attention flow network.

2.2   Attention Flow Network

Attention flow network is a directional weighted graph with attributes of nodes. It constructed by the website sequence generated by users' clickstream. Where nodes are websites, edges are users' jumping from one website to another by clicking, weight on the edges are the times that users jump from one website to another, and the attributes on the nodes mean the time that users stay on the websites [11].

There are many common patterns and evolution laws of the world wide web have been found in attention flow network, such as dissipation law, gravity law, Heap's law and Kleiber's law [12, 13]. In [14], they showed the flow distance of website through a 20 dimensional Euclidean space, and finds that 20% of websites attracts 75% of attention flow. In [15], they found the nonlinear relationship $C_i \sim A_i^{\gamma} (\gamma < 1)$ between site influence C_i and site traffic A_i by studying the decentralized flow structure of clickstream. In [16] views, website as a life organization from the aspect of ecology, site influence C_i as a metabolic rate and users' residence time T_i as volume. As a result, they discovered that there is a sublinear relationship between them, namely Kleiber's law.

3   Data and Problem Description

3.1   Data

This paper selects the data of online users' behaviors provided by China Internet Network Information Center (CNNIC). In windows operating system, the data recording program checks the current focus window of users' computers every two seconds. If the results of the two checks are inconsistent, a record is added to describe the behavior. It records the start-up time and last shut-down time, process names, URL addresses, current tab handles, window handles, program names, program company names and so on. This paper randomly selects about 120 million records from 1000 users' clicked behaviors in a month to establish attention flow network. A part of one user's cleaned data is shown in Table 1.

Table 1   A part of one user's cleaned data

2012/8/17 6:39	['sougou.com']
2012/8/17 6:40	['sougou.com']
2012/8/17 6:40	['newegg.com']
2012/8/17 6:40	['newegg.com']
2012/8/17 6:40	['sougou.com']
2012/8/17 6:41	['yiqifa.com']
2012/8/17 6:41	['happigo.com']
2012/8/17 6:43	['sougou.com']
2012/8/17 6:43	['sougou.com']

3.2   The Construction of Individual Attention Flow Network (IAFN)

Let $G_K = (V_k, E_k, T_k, W_k)$ represent the $User_k$'s attention flow network. Where $V_k = \{v_1, v_2, \ldots, v_n\}$ is the set of websites that the user clicks. $E_k = \{e_{ij} | 1 \leq i \leq n, 1 \leq j \leq n\}$, where $e_{ij} = 1$ is that the $User_k$ jumps from website v_i to website v_j. $T_k = \{t_1, t_2, \ldots, t_n\}$, where t_n is the accumulated time that the $User_k$ stays on website v_n. $W_k = \{w_{ij} | 1 \leq i \leq n, 1 \leq j \leq n\}$, where w_{ij} means the weight of the edge e_{ij}. In other words, w_{ij} represents the times of clicked behaviors between website i and j. The IAFN of user A is shown in Fig. 1, and user B is shown in Fig. 2.

3.3   The Construction of Collective Attention Flow Network

Let $G = \{G_1, G_2, \ldots, G_n\} = (V, E, T, W)$ indicate The Collective Attention Flow Network (CAFN). Where $V = \{V_1 \cup V_2 \cup \ldots \cup V_k\}$ is the set of websites visited by all users. $E = \{E_1 \cup E_2 \cup \ldots \cup E_k\}, \forall e_{ij} \in E$, Where $e_{ij} = 1$ is that at least one of the k users has jumped from website v_i to v_j. Where $e_{ij} = 0$ is that none of the k users has jumped from website v_i to v_j. $T = \{\sum t_1, \sum t_2, \ldots, \sum t_n\}$, where $\sum t_n$ is the accumulated time of all users stay on the website. $W = \{\sum w_{ij} | 1 \leq i \leq n, 1 \leq j \leq n\}$, where $\sum w_{ij}$ are the times that all users jump from website v_i to v_j. The CAFN constructed by the IAFN of user A and B is shown in Fig. 3.

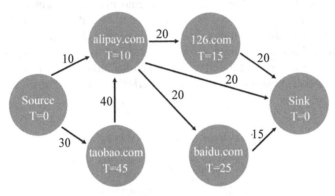

Fig. 1. The IAFN of user A

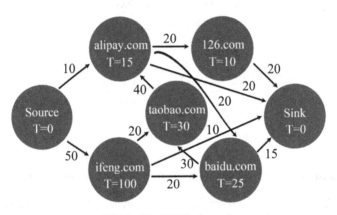

Fig. 2. The IAFN of user B

3.4 The Problem Description of Link Prediction in Attention Flow Networks

In IAFN of user A, there is no link between node baidu.com and node taobao.com. But it exists in CAFN. The existence of the link in CAFN is a necessary condition for the existence of the link in IAFN.

In formula (1), where V_k is the node set of IAFN of user k, T_k is the node stay time set of the IAFN of user k, W_k is the weight set of the IAFN of user k, and G is the CAFN. Then, the problem is equivalent to judging the G_k when the above conditions are known.

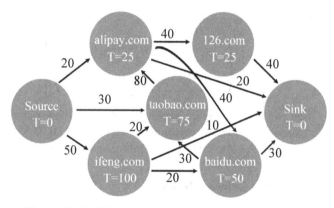

Fig. 3. The CAFN constructed by the IAFN of users A and B

The conditional probability of $e_{ij} = 1$ is as follows:

$$p_{ij} = P\left(e_{ij} = 1 | G, V_k, T_k, W_k\right) \tag{1}$$

The problem can be transformed by specifying a critical value ξ, as follows:

$$e_{ij} = \begin{cases} 1, p_{ij} \geq \xi, \\ 0, p_{ij} < \xi. \end{cases} \tag{2}$$

Finally, the link existence can be predicted in G_k by p_{ij}.

4   Feature Extraction

The structure of attention flow network is complex. For example, out degree, in degree, edge, weight, node attribute can be used as the feature input of the algorithm, but this will make the maximum entropy model too complex to be solved. Considering the scale-free nature of CAFN, this problem can be alleviated by the following four indexes.

4.1   Degree Influence Index

Nodes with big degrees are more likely to have edges. On the contrary, they are not. Degree influence index refers to the influence of degree on link existence probability. First, all nodes in CAFN are ranked in descending order according to the out-degrees and in-degrees. The num is the number of nodes. Then, for all nodes in IAFN of all users create the key-value pairs "<key: (user id, node), value: out-degree>" and "<key: (user id, node), value: in-degree>" according to the out-degree and in-degree. Then, these key-value pairs are ranked in descending order. The *knum* is the number of key-value pairs. If an edge e_{ij} in CAFN, then the out-degree of node i is $d_{c_i}^{out}$, the in-degree of node j is $d_{c_j}^{in}$. The out-degree of node i is $d_{k_i}^{out}$ and the in-degree of node j is $d_{k_j}^{in}$ in IAFN of user k. Then degree influence is as follows:

$$DI = \left\lfloor \frac{N}{4} \times \left(\frac{\frac{Rank\left(d_{c_i}^{out}\right)+Rank\left(d_{c_j}^{in}\right)}{num}}{+\frac{Rank\left((k,i),d_{k_i}^{out}\right)+Rank\left((k,j),d_{k_j}^{in}\right)}{knum}} \right) \right\rfloor \tag{3}$$

Where *Rank* is used to get the list index position by degree or key-value pairs sorted from largest to smallest. N is used to control the value range of DI. If N = 10, then it divides the influence of degree into 10 levels according to in-degrees and out-degrees.

4.2 The Edge and Weight Influence Index

If the weight of edges is big in CAFN, Then the possibility of edges existing in IAFN is big. The edge and weight influence index refer to the influence of edge and weight on link existence probability. First, all the edges in CAFN are ranked in descending order according to the weight. The num is the number of edges in CAFN. Next, for all the edges of IAFN create key-value pairs "<key: (user ID, out-node, in-node), value: weight of the edge>". Then rank them in descending order according to the weights. The *knum* is the number of key value pairs. The w_{ij}^c is the weight of the edge e_{ij} in CAFN. The w_{ij}^k is the weight of edge e_{ij} in IAFN of user k. Similarly, N controls the value range of EWI. Then the edge weight influence is as follows:

$$EWI = \begin{cases} \left\lfloor \frac{N}{2} \times \left(\frac{Rank\left(w_{ij}^c\right)}{num} + \frac{Rank\left((k,i,j),w_{ij}^k\right)}{knum} \right) \right\rfloor, & w_{ij}^k > 0, \\ \left\lfloor \frac{N}{2} \times \left(\frac{Rank\left(w_{ij}^c\right)}{num} + 1 \right) \right\rfloor, & w_{ij}^k = 0. \end{cases} \tag{4}$$

4.3 Influence of Staying Time Index

The longer the nodes staying time, the more likely the nodes have edges. Influence of staying time index refers to the influence of staying time on link existence probability. First, all nodes in CAFN are ranked in descending order of residence time. The num is the number of nodes. Then, for all nodes in attention flow network of all individual users create key-value pairs " <key: (user ID, node), value: stay time> ". Next, the key-value pairs are sorted according to the stay time. The *knum* is the number of key-value pairs. If an edge e_{ij} in CAFN, then in CAFN, the $t_{c_i}^{out}$ is the stay time of node i, the $t_{c_j}^{in}$ is the stay time of node j, and in IAFN of user k, the $t_{k_i}^{out}$ is the stay time of node i, the $t_{k_j}^{in}$ is the stay time of node j. Similarly, N controls the value range of IOST. The influence of stay time is as follows:

$$IOST = \left\lfloor \frac{N}{4} \times \left(\frac{\frac{Rank\left(t_{c_i}^{out}\right)+Rank\left(t_{c_j}^{in}\right)}{num}}{+\frac{Rank\left((k,i),t_{k_i}^{out}\right)+Rank\left((k,j),t_{k_j}^{in}\right)}{knum}} \right) \right\rfloor \tag{5}$$

4.4 The Individual Users' Network Influence Index

Different users have different levels of activity online. The more active users, the more likely to generate links. Therefore, the whole network structure should be considered. In IAFN of user k, the D_k is the sum of the in-degree and out-degree of all node, and the W_k is the weight of all edges, and the T_k is the sum of the stay time of all nodes. Then, for all user's IAFN create key-value pairs "<key: k, value: D_k> ", " <key: k, value: W_k> " and "<key: k, value: T_k>". Finally, rank the three key value sets in descending order. Where $knum$ is number of users. N controls the value range of UNI. The individual users' attention flow network is as follows:

$$\text{UNI} = \left\lfloor \frac{N}{3 \times \text{knum}} \times (\text{Rank}(k, D_k) + \text{Rank}(k, W_k) + \text{Rank}(k, T_k)) \right\rfloor \quad (6)$$

5 The Link Prediction Algorithm Based on Maximum Entropy Model

If there is a link e_{ij}^c in CAFN, then $e_{ij}^c = 1$. This link might exist in IAFN of user k, that is, $e_{ij}^k = 0$ or $e_{ij}^k = 1$. The problem of link prediction in IAFN of user k is, when $e_{ij}^c = 1$, then $e_{ij}^k = 0$ or $e_{ij}^k = 1$?

Choose any two nodes such as i and j in individual User$_k$'s attention flow network, then DI_{ij}^k, EWI_{ij}^k, $IOST_{ij}^k$, and UNI_{ij}^k will be found. Make these four indexs into a vector X, and the value of e_{ij}^k into y. The formula (1) can be rewritten as:

$$p_{ij} = P(y|X) \quad (7)$$

When P(y = 1|X) is greater than ξ, then y = 1, otherwise y = 0. The formula (2) can be rewritten as:

$$y = \begin{cases} 1, P(y = 1|X) \geq \xi, \\ 0, P(y = 1|X) < \xi. \end{cases} \quad (8)$$

Obtain the sample data set $\{(X_1, Y_1), (X_2, Y_2), \ldots, (X_i, Y_i)\}$ through the CAFN and IAFN. The characteristic function is introduced as follows:

$$f(x, y) = \begin{cases} 1, & \text{if } (x, y) \text{ in sample spaces}, \\ 0, & \text{other} \end{cases} \quad (9)$$

Where is $x \in X_i$, $y \in Y_i$. The empirical distribution expectation of f(x, y) is as follows:

$$E_{\breve{P}}(f) = \sum_{x,y} \breve{P}(x, y) f(x, y) \quad (10)$$

Where $\breve{P}(x, y)$ is as follows:

$$\breve{P}(x, y) = \frac{count(x, y) \ in \ sample \ \text{spaces}}{size \ of \ \text{sample spaces}} \quad (11)$$

The model estimate expectation of f(x, y) is as follows:

$$E_p(f) = \sum_{x,y} \breve{P}(x)p(y|X)f(x, y) \tag{12}$$

Make a constraint as follows:

$$E_p(f) = E_{\breve{p}}(f) \tag{13}$$

The definition of conditional entropy is as follows:

$$H(Y|X) = -\sum p(x, y) \log p(y|x) \tag{14}$$

It is transformed into an optimal problem satisfying a set of constraints as follows:

$$P = \left\{ p(y|X)|E_p(f_i) = E_{\breve{p}}(f_i) \right\},$$
$$p^* = \text{arg}maxH(Y|X) \tag{15}$$

This model can be solved by using the Lagrange multiplier method. Then the probability p^* of maximum entropy under the condition $E_p(f_i) = E_p(f_i)$ can be obtained as follows:

$$p^*(y|X) = \frac{exp(\sum_i \lambda_i f_i(x,y))}{Z(x)} \tag{16}$$

Where λ_i is the weight factor, and $Z(x)$ is the normalization factor. So far, the maximum entropy model has been established. For detailed derivation, please refer to reference [17]. Then, the conditional probability shown in formula (7) can be rewritten as follows:

$$p(y|X) = p^*(y|X) = \frac{exp(\sum_i \lambda_i f_i(x,y))}{Z(x)} \tag{17}$$

The common algorithms for solving the maximum entropy model are IIS, GIS [18], L-BFGS [19], etc. Compared with GIS algorithm, the implementation of IIS and L-BFGS algorithm is a little more complex. They are usually used to solve large-scale problems. In this paper, the scale of the problem is moderate, so GIS algorithm is used to solve the maximum entropy model. The GIS algorithm first initializes all eigenvalues to 0 as follows:

$$\lambda_i^{(0)} = 0, i \in \{1, 2, \ldots, n\} \tag{18}$$

Where n is the number of characteristic functions. Then iterate as follows:

$$\lambda_i^{(itor+1)} = \lambda_i^{(itor)} + \frac{1}{C} \log \frac{E_p(f_i)}{E_{\tilde{p}}(f_i)} \tag{19}$$

Where itor is the times of iterations; C is maximum of the number of elements in set X_i in the sample space. There are $C = |X_i| = 4$ for any sample data. With the increasing times of iterations, the empirical distribution expectation $E_p(f_i)$ of characteristic function f_i will be equal to that of its model estimation $E_{\breve{p}}(f_i)$. Until satisfied conditions as follows:

$$\lambda_i^{(itor+1)} - \lambda_i^{(itor)} \leq \delta, i \in \{1, 2, \ldots, n\} \tag{20}$$

Where δ is the convergence threshold. When difference between the λ_i of the last two iterations is less than or equal to the δ, the training of λ_i is completed.

After all the λ_i are solved by GIS algorithm, the $p(y = 1|X)$ can be computed from formula (17). Then formula (8) can be used to determine whether $e_{ij}^k = 0$ or $e_{ij}^k = 1$ in the IAFN of user k, when $e_{ij}^c = 1$ in the CAFN. The implementation of MELP algorithm is shown in Algorithm 1.

Algorithm 1: MELP

Input:
　　Maximum number of iterations N
　　Sample data set S
　　Convergence threshold δ
Output:
　　Weight $\lambda_1, \lambda_2, \ldots, \lambda_i$

```
1:   Initialize featureList as dictionary objects
2:   for t ∈ (DI, EWI, IOST, UNI) in S:
3:       for label in (1,0):
4:           key=string(t) concat string(label)
5:           if key in featureList: featureList[key]+=1;
6:           else: featureList[key]=0;
7:   Initialize weight, empiricalE
8:   for key in featureList:
9:       weight[key]=0
10:      empirical[key]=featureList[key]/length(S)
11:  for i in range(N):
12:      Initialize modelE as dictionary objects
13:      for j in range(length(S)):
14:          calculate p₀ and p₁ by formula (17), when label=0 and label=1
15:      for key in (DI, EWI, IOST, UNI):
16:          for label in (1,0):
17:              key=string(t) concat string(label)
18:              modelE[key]+=p[label]*(1.0/length(S))
19:      lastWeight=weight
20:      for key in featureList:
21:          weight[key]+=(1.0/C)*log(empiricalE[key]/ modelE[key])//C=4
22:      flag=true
23:      for key in featureList:
24:          if lastWeight[key]-weight[key]> δ:
25:              flag=false; break;
26:      if flag==true: break;
27: output weight=λ₁, λ₂, ... , λᵢ
```

---

6 Experimental Analysis

Select the cleaned data in Sect. 3.1 to construct the network. The constructed CAFN has 20,746 nodes and 136,448 edges, as shown in Fig. 4. This paper uses support vector machine (SVM), perceptron (Pre) and naive Bayes (NBC) to carry out comparative experiments.

6.1 Algorithms Complexity Analysis

In Algorithm 1, the time complexity of steps 1 to 6 is $O(n)$, and steps 7 to 10 is $O(f)$, and steps 11 to 27 is $O(i*n)$, So we can calculate that the time complexity of MELP is $O(f + i*n)$. Where n is the size of sample spaces, and f is the number of features, and i is the maximum number of iterations. Usually, the time complexity of SVM is $O(n^2)$, and PRE is $O(n*d*i)$, and NBC is $O(n*d)$. Where d is feature dimension. The order of time complexity from the best to the worst is as follows: NBC, MELP, PRE and SVM. In fact, the time complexity of MELP is approximately equal to NBC in this problem, and NBC has a great disadvantage in other evaluation indexes.

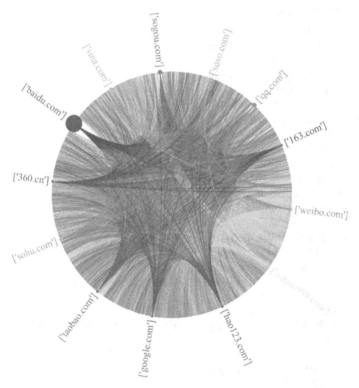

Fig. 4. The collective attention flow network

6.2 Evaluate Indexes by Machine Learning Algorithm

Tenfold cross validation is used to train and predict the algorithm. The predicted results are compared on the four indexes: precision (Prec), accuracy (Acc), recall rate (Rec), and F1 value. The results are shown in Fig. 5.

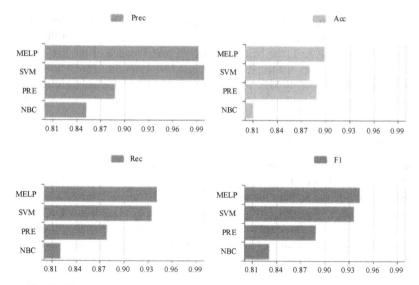

Fig. 5. The comparison on prediction performance among different algorithms

MELP algorithm is superior to Pre and Nbc algorithm in precision, accuracy, recall and F1 value. SVM algorithm has no obvious advantage in accuracy index. Compared with SVM in the above four indicators, the overall performance of MELP is better.

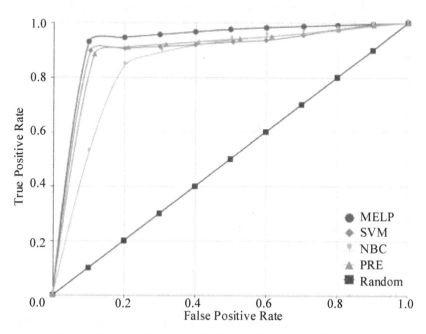

Fig. 6. The prediction results of different algorithms in ROC curve

The comparison results of ROC curve are shown in Fig. 6. The closer the ROC curve is to the upper left corner, the better the overall prediction effect of the algorithm is. The closer the AUC value is to 1, the better the overall performance of the link prediction algorithm. The AUC values of MELP, SVM, Pre and Nbc algorithm are 0.9686, 0.9369, 0.9310 and 0.8518. Obviously MELP is superior to the other three algorithms in ROC curve and AUC value.

6.3 Evaluate Indexes by Link Prediction Algorithm

Lü and Zhou proposed AUC index to evaluate the link prediction algorithm in paper [8]. The data in the test set is divided into two sets: E_1 and E_2. The edge E_1 in the set exists in the network while the edge in the set E_2 does not exist. In each random experiment, one edge e_1 is randomly put back from E_1, and one edge e_2 is randomly put back from E_2. Let $n_1 = 0$, $n_2 = 0$, if the score value of the edge e_1 is higher than that of edge e_2, n_1 will increase 1; if the score value of the edge e_1 is equal to that edge e_2, n_2 will increase by 1. Repeat this random experiment N times, and the AUC value of the link prediction algorithm can be obtained as follows:

$$\text{AUC} = \frac{n_1 + 0.5 * n_2}{N} \tag{21}$$

The sample score value of SVM algorithm and Pre algorithm is defined as the distance from the sample point to the hyperplane. The sample score value of MELP and Nbc algorithm is the probability value of the link. The times of repeated random experiments is the number of samples in the test set, and the experimental results are shown in Fig. 7. Obviously, MELP algorithm has an advantage in this AUC value.

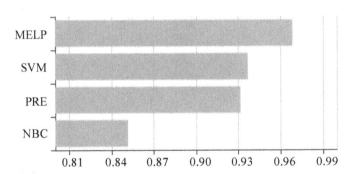

Fig. 7. Comparison on AUC values corresponding to different algorithms

7 Conclusion

MELP algorithm is put forward according to the advantages of the maximum entropy model without extra conditions and independent assumption between variables. Experiments show that: 1) the MELP algorithm proposed in this paper is superior to the perceptron (Pre) and naive Bayes (Nbc) in accuracy, recall rate and F1 value. 2) the overall performance is superior to the support vector machine (SVM) algorithm; 3) the MELP algorithm is superior to the comparison algorithm on AUC value and ROC curve. It is the first time to study link prediction in the field of attention flow network research. MELP algorithm can be applied to online advertising, perception of users' risky behavior and other fields. MELP algorithm also has reference significance for solving link prediction problems of other network models.

The advantages of MELP algorithm are as follows: 1) due to without extra independent assumptions and constraints, the classification results are more accurate; 2) the unbiased principle is adopted to ensure the uniform distribution of unknown information, which has more advantages for the classification of unbalanced sample data. However, MELP algorithm also has some shortcomings: 1) with the increasing of sample data and dimensions, the number of features grows too fast, which will lead to the problem unsolved; 2) the existing solving algorithms are not perfect, because the GIS algorithm cannot solve the problem of large-scale model, and L-BFGS and IIS easily fall into local optimal solution.

In the future, the structure of attention flow network will be further analyzed to expand the dimensions of sample data reasonably. Some feature selection algorithms will be used to select some features that have a greater impact on the prediction results to train the algorithm. All these measures will be taken to improve the performance of MELP algorithm.

Acknowledgements. This research was partially supported by the grants from the Natural Science Foundation of China (No.71764025,61863032); the Research Project on Educational Science Planning of Gansu, China (Grant No. GS[2018]GHBBKZ021, GS[2018]GHBBKW007). Author contributions: Yong Li and Jingpeng Wu are co-first authors who jointly designed the research.

References

1. Watts, D.J., Strogatz, S.H.: Collective dynamics of 'small-world' networks. Nature **393**(6684), 440 (1998)
2. Barabási, A.L., Albert, R.: Emergence of scaling in random networks. Science **286**(5439), 509–512 (1999)
3. Barabási, A.L.: Network science: luck or reason. Nature **489**(7417), 507 (2012)
4. Getoor, L., Diehl, C.P.: Link mining. ACM SIGKDD Explor. Newslett. **7**(2), 3–12 (2005)
5. Li, Y.J., Yin, C., Yu, H., Liu, Z.: Link prediction in microblog retweet network based on maximum entropy model. Acta Phys. Sin. **65**(2), 020501 (2016). (in Chinese)
6. Wu, S., Sun, J., Tang, J.: Patent partner recommendation in enterprise social networks. In: Proceedings of the sixth ACM International Conference on Web Search and Data Mining. ACM (2013)
7. Li, Y.Q., Chen, W.Z., Yan, H.F., Li, X.M.: Learning graph-based embedding for personalized product recommendation. Chin. J. Comput. **42**(8), 1767–1778 (2019). (in Chinese)
8. Lü, L., Zhou, T.: Link prediction in complex networks: a survey. Phys. A Stat. Mech. Appl. **390**(6), 1150–1170 (2011)
9. Hou, W., Huang, Y., Zhang, K.: Research of micro-blog diffusion effect based on analysis of retweet behavior. In: IEEE International Conference on Cognitive Informatics & Cognitive Computing, pp. 255–261. IEEE (2015)

10. Liu, W., He, M., Wang, L.H., Liu, Y., Shen, H.W., Cheng, X.Q.: Research on microblog retweeting prediction based on user behavior features. Chin. J. Comput. **39**(10), 1992–2006 (2016). (in Chinese)

11. Lou, X., Li, Y., Gu, W., Zhang, J.: The atlas of chinese world wide web ecosystem shaped by the collective attention flows. PLoS ONE **11**(11), e0165240 (2016)

12. Li, Y., Meng, X.F., Zhang, Q., Zhang, J., Wang, C.Q.: Common patterns of online collective attention flow. Sci. China Inf. Sci. **60**(5), 59102:1–059102:3 (2017). https://doi.org/10.1007/s11432-015-0567

13. Wu, F., Huberman, B.A.: Novelty and collective attention. Proc. Natl. Acad. Sci. U.S.A. **104**(45), 17599–17601 (2007)

14. Shi, P., Huang, X.H., Wang, J., Zhang, J., Deng, S., Wu, Y.H.: A geometric representation of collective attention flows. PLoS ONE **10**(9), e0136243 (2015)

15. Wu, L., Zhang, J.: The decentralized flow structure of clickstreams on the web. Eur. Phys. J. B **86**(6), 266 (2013). https://doi.org/10.1140/epjb/e2013-40132-2

16. Li, Y., Zhang, J., Meng, X.F., Wang, C.Q.: Quantifying the influence of websites based on online collective attention flow. J. Comput. Sci. Technol. **30**(6), 1175–1187 (2015). https://doi.org/10.1007/s11390-015-1592-4

17. Berger, A.L.: A maximum entropy approach to natural language processing. Comput. Linguist. **22**(1), 39–71 (1996)

18. Darroch, J.N., Ratcliff, D.: Generalized iterative scaling for log-linear models. Ann. Math. Stat. **43**(5), 1470–1480 (1972)

19. Byrd, R.H., Nocedal, J., Schnabel, R.B.: Representations of quasi-Newton matrices and their use in limited memory methods. Math. Program. **63**(1–3), 129–156 (1994). https://doi.org/10.1007/BF01582063

Graph Representation Learning Using Attention Network

Bijay Gaudel[1], Donghai Guan[1(✉)], Weiwei Yuan[1], Deepanjal Shrestha[1], Bing Chen[1], and Yaofeng Tu[2]

[1] College of Computer Science and Technology, Nanjing University of Aeronautics and Astronautics, Nanjing, China
{dhguan,yuanweiweig,cb_china}@nuaa.edu.cn
[2] ZTE Corporation, Shenzhen, China
tu.yaofeng@zte.com.cn

Abstract. Network embedding is a method to learn low dimensional representation of nodes in large graph with the goal of capturing and preserving the network structure. Graph convolution networks (GCN) are successfully applied in node embedding task as they can learn sparse and discrete dependency in the data. Most of the existing work in GCN requires costly matrix operation. In this paper, we proposed a graph neighbor Sampling, Aggregation, and ATtention (GSAAT) framework. That does not need to know the graph structure upfront and avoid costly matrix operation. The proposed method first learn to aggregate the information of node's neighbors and stacked a layer in which nodes are able to attend over the aggregated information of their neighbors feature. The proposed method achieved state-of-art performance in two classification benchmark: Cora, and Citeseer citation network dataset.

Keywords: Node embedding · Graph convolution networks · Graph attention

1 Introduction

Large information networks are ubiquitous in real world with the example such as scientific collaboration networks, social networks, and biological networks. Low dimensional embedding of large network nodes proves to be crucial for many network analysis task. The main idea behind the node embedding is to learn low dimensional vector representation of nodes with the goal of reconstructing the network in learned embedding space. Tasks such as Link prediction [1], node classification [2–5], information retrieval [6] and clustering [7] are done directly by feeding node's low dimensional vector representation as feature into the downstream machine learning algorithms.

Most of the early works [1,2,5,7–10] are either based on random walk on the network or matrix factorization. Matrix factorization based methods cannot generalize the network with different structure. Random walk based methods

© Springer Nature Singapore Pte Ltd. 2021
H. Mei et al. (Eds.): BigData 2020, CCIS 1320, pp. 137–147, 2021.
https://doi.org/10.1007/978-981-16-0705-9_10

convert the network structure into sequence and preserve the proximity based on co-occurrence statics in the sequence. These methods are unsupervised and can not perform end-to-end node classification task. Several research work in graph convolution neural networks (GNNs) have been carried out in recent years to perform node classifications task in end-to-end fashion such as [11,20,24,27,28].

Convolution neural network (CNN) have been successfully applied in the filed such as machine translation [11], image classification [12], speech recognition [13] where the data is in euclidean domain. But these methods fail to adress social networks, biological networks, telecommunication networks, where data exist in irregular domain in graph structure rather than the euclidean domain [14–16]. Few approaches [17–20] have been carried out for adopting spectral approaches to desire CNN however those approaches cannot be applicable to networks with different structure because they depends on graph laplacian eigenbasis. On the other hand, many non-spectral approaches [15,21–23,27], define convolution on graph (network) by directly on group of spatially close neighbors. The main goal of non-spectral approach is to define the convolution on the graph which works directly without defining the convolution in Fourier domain.

Attention mechanism is highly used in sequence based task [11,24]. The main goal of attention mechanism is to focus more on important features and less on unimportant features. Attention mechanism helps to improve the method based on recurrent neural network (RNN) and CNN. Recently self attention approaches [25,28] in graph neural network have been studies. Graph Attention Network (GAT) [28] is based on stacking a node over neighbor features, which improves performance on graph related task. Attention using graph convolution helps to consider both nodes feature and graph topology.

In this paper, we propose a simple transductive framework called GSAAT to perform node embedding of the network. The main idea of our proposed framework is to sample fixed number of neighbor for the nodes and aggregate the features of neighbor. And then, perform nodes features attention over aggregated features of neighbors. The Proposed method is based on transductive approach which learns the embedding function in non-euclidean domain. The Proposed method is computationally efficient as the aggregation and attention operation can be parallelizable across every node-neighbor pair. We evaluate our proposed method on two real world benchmark: Cora and Citesser. Using these benchmark we show that our proposed method achieve the state-of-art performance with 2.89% and 3.86% accuracy improvement in Cora and Citesser citation network data respectively.

2 Related Work

In the past few years, many graph embedding methods that learn low-dimensional embedding representation based on matrix factorization and random walk statistics [1–5,8] have been studied widely. These embedding methods directly train individual node to learn embedding representation of a node. They are inherently transductive and can not generalize to unseen node, and need expensive additional stochastic gradient descent training to make prediction on unseen nodes.

2.1 Graph Convolution

Graph neural networks (GNN) is the *de facto* standard in graph representation task for the for semi-supervised approach. Several attempts have been made to use graph convolution network for the node embedding in the literature. So far the convolution operator on graph are either based on spectral or non-spectral domain. In [18], convolution is defined in fourier domain, but this method is computationally expensive and convolution filters are non-spatially localized. Issue of non-spatiality is solved by work presented in [18] by introducing a parameterization of the spectral filters to make spatially localized filter. Work in [19] define the layer wise propagation rule using Chebyshev expansion in graph laplacian method and removed the necessity to calculate eigenvector of laplacian. The method presented in [20] restricted the convolution filter to operate around only first-hop neighbors. All these mentioned methods are based on spectral approach and need to calculate laplacian eigen basis. Works presented in [15, 21–23, 27] are based on non-spectral approach. The method in [23] is based on spatial approach and do unified generalization of CNN over the network. The approach in [27] is learning node embedding by sampling and aggregation of fixed number of nodes-neighbor pairs. These methods have significantly improved the classification task in the network.

2.2 Graph Attention

Recently attention based mechanism has captivated intrest in graph representation task. The main benifit of attention mechanism is, they allowed variable size input and focus on important part predict. Methods in [25, 28] are based on self attention over the graph. The idea presented in [28] is to stack the self masked attention layer over nodes so that they can attend over their neighbors. The approach in [25] to apply pooling method based on self attention which allows this method to consider both node feature and graph topology.

Our model draws inspiration from these previous work of graph convolution neural network, graph attention, and feature aggregation in network. We have build new approach for graph representation on the basis of Aggregation and Attention over graph.

3 Proposed Model: GSAAT

Inspired from inductive representation learning on large graph [27], and graph attention network [28], we proposed a novel neural network architecture GSAAT. The key idea behind GSAAT is learn to aggregate feature information from node's local neighborhood and stack layers in which nodes are able to attend over their local neighborhood aggregated features.

Table 1. Summary of datasets

	Cora	Citeseers
#Nodes	2708	3327
#Edges	5429	4732
#Class	7	6
#Features/node	1433	3703
#Training nodes	140	120
#Validation nodes	500	500
#Test nodes	1000	1000

3.1 Aggregator Layer

Graph data can not be represented in a grid like structure, and nodes neighbor have no natural ordering. The aggregator function should not be affected by the randomness of neighbor ordering. Our neural network model can be trained with randomly sampled nodes local neighborhood feature sets to maintain high representational capacity of the model. In practice, we examine three candidate aggregator functions presented in the paper [27]. The aggregator function in our proposed method is slightly different from presented in [27]. Unlike in [27] we only aggregate neighbor features and stack the nodes features over aggregated neighbor features for attention mechanism.

Mean Aggregator: The mean aggregator that we used is similar to convolution propagation rule in transductive GCN method [20]. This mean aggregator simply takes elementwise mean of the vectors in $\{h_u^k, \forall u \in N(v)\}$

Pooling Aggregator: In this aggregator neighbor vector is transformed by fully connected single neural network and either elementwise max-pooling or elementwise mean-pooling is performed to aggregate information across the nodes neighbor.

$$\text{AGGREGATE}_k^{pool} = max((W_{pool}h_u^k + b)\forall u \in N(u))$$

or

$$\text{AGGREGATE}_k^{pool} = mean((W_{pool}h_u^k + b)\forall u \in N(u))$$

where max, $mean$, u, and $N(u)$ represents the element-wise maximum and element-wise minimum operator nodes and neighbor of nodes respectively. Unlike in [27] we do not use activation function in this aggregator layer. Pooling aggregator scales down the size of representation results into less number of parameter for convolution network, which helps to avoids the over-fitting problem.

LSTM Aggregator: This aggregator has a higher representational capacity than both mean and pooling aggregator. Inherently LSTM are not permutation Inherent but here we adopt LSTM aggregator as permutation-invariant function.

We train this aggregator by shuffling the node's neighbors order, which is enough to teach the LSTM to ignore the sequence ordering.

3.2 Attention Layer

Influenced by the self attention layer architecture in [28] we designed nodes attention over aggregated information of nodes neighbor. Input to the attention layer is node features and aggregated neighborhood features, $h = \{\vec{h_1}, \vec{h_2}, \cdots, \vec{h_N}\}, \vec{h_i} \in \mathbb{R}^F$ and $b = \{\vec{b_1}, \vec{b_2}, \cdots, \vec{b_N}\}, \vec{b_i} \in \mathbb{R}^F$ respectively. Where, N is the number of nodes and F is the number of features of nodes and aggregated neighborhood features. The attention layer produces the new set of node features by stacking nodes feature over aggregated neighborhood features. The output of the attention layer is the new set of nodes feature with different cardinality F', $h' = \{\vec{h_1}', \vec{h_2}', \cdots, \vec{h_N}'\}, \vec{h_i}' \in \mathbb{R}^{F'}$.

We applied shared linear transformation parameterized by a weight matrix $W \in \mathbb{R}^{F' \times F}$ in both nodes features and aggregated neighbor features, in order to achieve high expressive power. Later we performed nodes attention over aggregated neighborhood information. The attention mechanism is denoted by $a : \mathbb{R}^{F'} \times \mathbb{R}^{F'} \to \mathbb{R}$; computes attention coefficients.

$$e_{ij} = a(W\vec{h_i}, W\vec{b_j}) \tag{1}$$

where, $j \in N(i)$. We normalize attention coefficients using softmax function, so that the coefficients become easily comparable across all the neighbor aggregated features.

$$\alpha_{ij} = softmax(e_{ij}) = \frac{exp(e_{ij})}{\sum_{k=N(i)}(exp(e_{ik}))} \tag{2}$$

We used single layer feed-forward neural network with weight vector $\vec{a} \in \mathbb{R}^{2F'}$ and applied LeakyReLU non-linearity with $\alpha = 0.2$ as in [28] and can be expressed as:

$$\alpha_{ij} = \frac{exp(LeakyReLU(\vec{a}^T[W\vec{h_i}\|W\vec{b_j}]))}{\sum_{k\in N(i)} exp(LeakyReLU(\vec{a}^T[W\vec{h_i}\|W\vec{b_k}]))} \tag{3}$$

where $\|$ denotes concatenation operation and ^{T} represents transposition. After Calculating normalized attention coefficients, linear combination of features corresponding to them can be calculated. We used multihead attention mechanism to stabilize the learning process as in [28]. The output feature representation can be represented as:

$$\vec{h_i}' = \|_{k=1}^k \sigma\left(\sum_{j\in N(i)} \alpha_{ij}^k W^k \vec{h_j}\right) \tag{4}$$

Where $\|$ is concatenation, σ is a sigmoid function applied to achieve non linear output, α_{ij}^k are normalized attention coefficient by k^{th} attention (a^k) and W^k is the corresponding linear transformation weight matrix.

For the final layer prediction output we used single-head attention mechanism.

4   Experiments

We evaluate our proposed method in two benchmark datasets: *(i)* Cora, and *(ii)* Citeseers. We shows the effectiveness of our proposed model over the popular baseline methods in node classification task on these benchmark.

4.1   Datasets

We use two standard benchmark *(i)* Cora, and *(ii)* Citeseers dataset [26]. In both datasets nodes represents documents and edges between nodes represents citation. Both of the network are undirected. The Cora datasets have 2708 nodes, 5429 edges and Citeseers dataset have 3327 nodes, 44324 edges. All the nodes in both datasets are labeled into particular classes. Cora datasets nodes are labeled into 7 different classes and Citeseers dataset nodes are labeled into 6 different classes. Single node of Cora dataset have 1433 features per node and 3703 features per node for Citeseers dataset. For these datasets we allow only 20 nodes per classes for the training. Further we use 500 nodes for validation during the training time and 1000 nodes for testing our model. The summary of datasets, that we used to verify our model, is presented in Table 1.

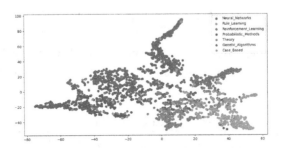

Fig. 1. Visualization of prediction of Cora dataset by GSAAT-mean method in embedded space.

4.2   Baseline Methods

We use following state-of-art baseline methods to compare the result with our proposed GSAAT methods.

ManiReg [29]. It is based on new type of regularization. It works on transductive as well as semi-supervised settings.

Semiemb [30]. It is a non linear embedding algorithm based on semi-supervised approach.

Deepwalk [2]. This method generate embedding via random walks in the network. Nodes are classified by feeding the embedding into SVM classifier.

LP [31]. Embedding algorithm based on continuous relaxation on the Markov random fields.

planteiod [30]. This method learns node embedding able to predict the context in network directly.

Chebyshev [19]. This algorithm is based on fast localized filters on the graph.

GCN [20]. It is based on layer wise propagation rule which operate directly over the graph. It scales linearly in the number of graph edges and node features.

GAT [28]. It is based on masked self attention mechanism and can directly apply over the graph. This method does not require to know the graph structure. It can be directly apply to transductive and inductive settings.

4.3   Experimental Setup

We applied two layer GSAAT model with sample number of nodes equal to 10. First layer consist of 8 attention head computing 8 features (total number of 64 features). Final layer is designed with single attention head for multi-label classification output. We used adam optimizer with learning rate 0.005. We have applied L2-regularization with $\lambda = 2.5e - 6$, and dropout regularization with $p = 0.4$ in both layers as well as normalized attention coefficient. We trained our model with EarlyStopping callback with 1000 epochs. Patience of 100 is applied in EarlyStopping callbacks by monitoring validation accuracy. We reproduced result for all the baseline from the paper [28].

5   Results

We report the mean classification accuracy by GSAAT, and all the other baseline methods on the test nodes of the datasets in Table 2. First row in this table represents datasets and first column represents GSAAT and all the other baseline methods. Each cell of the table presents the result of corresponding method in the corresponding datasets. Both *max*, and *mean* pooling aggregator give same result so we use *max* pooling aggregator in the experiment. Highest value in each column is presented in bold text. Our method is supervised so we compare all the results with baseline methods in supervised settings. We run all the methods for 10 times and report the mean result in Table 2. Results shows our method successfully achieved or matched state-of-art results in both datasets. Our method improves result by 2.89% in Cora dataset and 3.86% in Citeseer dataset compared to that of best performing baseline methods. Effectiveness of our method to learn Cora dataset to represent nodes features in embedding

Table 2. Summary of results interms of mean accuracy on Cora and Citeseers dataset.

Methods name	*Cora*	*Citeseers*
ManiReg	59.5%	60.1%
SemiEmb	59.0%	59.6%
Deepwalk	67.2%	43.2%
ICA	75.1%	69.1%
Planetoid	75.7%	64.7%
Chebyshev	81.2%	69.8%
GCN	81.5%	70.3%
GAT	$83.0 \pm 0.7\%$	$72.5 \pm 0.7\%$
GSAAT-mean (ours)	84.2%	74.8%
GSAAT-pool (ours)	**85.4%**	74.9%
GSAAT-LSTM (ours)	85.2%	**75.3%**

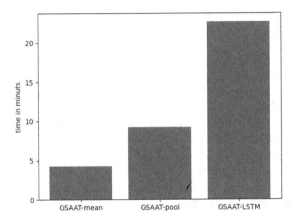

Fig. 2. Running time for proposed GSAAT methods with different aggregator.

space can be visualize in Fig. 1. We also performed running time analysis for The proposed methods with different aggregator as shown in Fig. 2 for Cora dataset. Our mean aggregator has constant time complexity hence it runs almost as fast as other baseline methods with higher accuracy in node classification task compared to baseline methods. We reported the effect of number of neighbor sampling in our methods with mean aggregator in Fig. 3. Cora dataset is used for this experiment. Our experiment shows sampling higher number of neighbor does not help to improve the effectiveness of our model significantly rather it becomes computationally expensive as shown in Fig. 4.

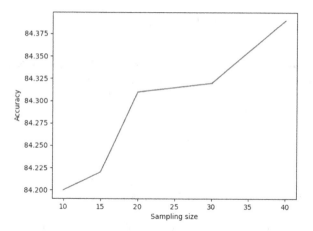

Fig. 3. Effect of sampling size on classification accuracy.

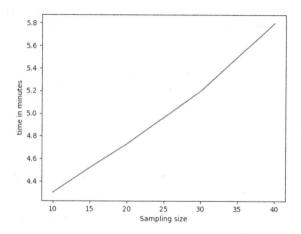

Fig. 4. Effect of sampling size on training+testing time.

6   Conclusion

In this work we proposed a new architecture of convolution neural network namely GSAAT. which learns to aggregate the node's neighbor information which helps to learn the graph structure as well as nodes features. After aggregating neighbor features, the proposed method do node's feature attention over aggregated features. The proposed GSAAT method is used to perform an experiment on two different benchmark: Cora and Citeseer to show the effectiveness of our proposed method. The proposed GSAAT method improves performance by 2.89% in Cora and 3.86% in Citeseer dataset. We have analyzed the effect of sampling size of neighbors in accuracy and running time. From the experiment the GSAAT with mean aggregator shows best result in terms of running time whereas GSAAT with pooling aggregator gives the best result in terms of accuracy.

Acknowledgements. Author like to thank Samadhan Engineering, Nepal for providing resources and suggestion for the improvement of result.

This work was supported by the Key Research and Development Program of Jiangsu Province (BE2019012).

References

1. Grover, A., Leskovec, J.: node2vec: scalable feature learning for networks. In: Proceedings of the 22nd ACM SIGKDD International Conference on Knowledge Discovery and Data Mining (2016)
2. Bryan, P., Al-Rfou, R., Skiena, S.: DeepWalk: online learning of social representations. In: Proceedings of the 20th ACM SIGKDD International Conference on Knowledge Discovery and Data Mining (2014)
3. Wang, D., Cui, P., Zhu, W.: Structural deep network embedding. In: Proceedings of the 22nd ACM SIGKDD International Conference on Knowledge Discovery and Data Mining (2016)
4. Tang, J., et al.: LINE: large-scale information network embedding. In: Proceedings of the 24th International Conference on World Wide Web (2015)
5. Ribeiro, L.F.R., Saverese, P.H.P., Figueiredo, D.R.: struc2vec: learning node representations from structural identity. In: Proceedings of the 23rd ACM SIGKDD International Conference on Knowledge Discovery and Data Mining (2017)
6. Weiss, Y., Torralba, A., Fergus, R.: Spectral hashing. In: Advances in Neural Information Processing Systems (2009)
7. Ng, A.Y., Jordan, M.I., Weiss, Y.: On spectral clustering: analysis and an algorithm. In: Advances in Neural Information Processing Systems (2002)
8. Cao, S., Lu, W., Xu, Q.: GraRep: learning graph representations with global structural information. In: KDD (2015)
9. Wang, X., Cui, P., Wang, J., Pei, J., Zhu, W., Yang, S.: Community preserving network embedding. In: AAAI (2017)
10. Xu, L., Wei, X., Cao, J., Yu, P.S.: Embedding identity and interest for social networks. In: WWW (2017)
11. Gehring, J., Auli, M., Grangier, D., Dauphin, Y.N.: A convolutional encoder model for neural machine translation. CoRR, abs/1611.02344 (2016). http://arxiv.org/abs/1611.02344
12. He, K., Zhang, X., Ren, S., Sun, J.: Deep residual learning for image recognition. In: Proceedings of the IEEE Conference on Computer Vision and Pattern Recognition, pp. 770–778 (2016)
13. Hinton, G., et al.: Deep neural networks for acoustic modeling in speech recognition: the shared views of four research groups. IEEE Sig. Process. Mag. **29**(6), 82–97 (2012)
14. Lazer, D., et al.: Life in the network: the coming age of computational social science. Science **323**(5915), 721 (2009)
15. Duvenaud, D.K., et al.: Convolutional networks on graphs for learning molecular fingerprints. In: Advances in Neural Information Processing Systems, pp. 2224–2232 (2015)

16. Davidson, E.H., et al.: A genomic regulatory network for development. Science **295**(5560), 1669–1678 (2002)

17. Bruna, J., Zaremba, W., Szlam, A., LeCun, Y.: Spectral networks and locally connected networks on graphs. In: International Conference on Learning Representations (ICLR) (2014)

18. Henaff, M., Bruna, J., LeCun, Y.: Deep convolutional networks on graph-structured data. arXiv preprint arXiv:1506.05163 (2015)

19. Defferrard, M., Bresson, X., Vandergheynst, P.: Convolutional neural networks on graphs with fast localized spectral filtering. In: Advances in Neural Information Processing Systems, pp. 3844–3852 (2016)

20. Kipf, T.N., Welling, M.: Semi-supervised classification with graph convolutional networks. In: International Conference on Learning Representations (ICLR) (2017)

21. Atwood, J., Towsley, D.: Diffusion-convolutional neural networks. In: Advances in Neural Information Processing Systems, pp. 1993–2001 (2016)

22. Niepert, M., Ahmed, M., Kutzkov, K.: Learning convolutional neural networks for graphs. In: Proceedings of the 33rd International Conference on Machine Learning, vol. 48, pp. 2014–2023 (2016)

23. Monti, F., Boscaini, D., Masci, J., Rodola, E., Svoboda, J., Bronstein, M.M.: Geometric deep learning on graphs and manifolds using mixture model CNNs. arXiv preprint arXiv:1611.08402 (2016)

24. Bahdanau, D., Cho, K., Bengio, Y.: Neural machine translation by jointly learning to align and translate. In: International Conference on Learning Representations (ICLR) (2015)

25. Lee, J., Lee, I., Kang, J.: Self-attention graph pooling. In: International Conference on Machine Learning (2019)

26. Sen, P., Namata, G., Bilgic, M., Getoor, L., Galligher, B., Eliassi-Rad, T.: Collective classification in network data. AI Mag. **29**(3), 93 (2008)

27. Hamilton, W., Ying, R., Leskovec, J.: Inductive representation learning on large graphs. In: Advances in Neural Information Processing Systems (2017)

28. Veličković, P., et al.: Graph attention networks. In: International Conference on Learning Representations (2018)

29. Belkin, M., Niyogi, P., Sindhwani, V.: Manifold regularization: a geometric framework for learning from labeled and unlabeled examples. J. Mach. Learn. Res. **7**, 2399–2434 (2006)

30. Weston, J., Ratle, F., Mobahi, H., Collobert, R.: Deep learning via semi-supervised embedding. In: Montavon, G., Orr, G.B., Müller, K.-R. (eds.) Neural Networks: Tricks of the Trade. LNCS, vol. 7700, pp. 639–655. Springer, Heidelberg (2012). https://doi.org/10.1007/978-3-642-35289-8_34

31. Weston, J., Chopra, S., Bordes, A.: Memory networks. CoRR, abs/1410.3916 (2014). http://arxiv.org/abs/1410.3916

Food Pairing Based on Generative Adversarial Networks

Yan Bai[2], Chuitian Rong[1,2]([envelope]) [iD], and Xiangling Zhang[3]

[1] Tianjin Key Laboratory of Autonomous Intelligence Technology and Systems, Tianjin, China
[2] School of Computer Science and Technology, Tiangong University, Tianjin, China
{1831125447,chuitian}@tiangong.edu.cn
[3] Beijing Institute of Education, Beijing, China
zhangxiangling@bnu.edu.cn

Abstract. The Generative Adversarial Networks (GAN) has received great attention and achieved great successes in many applications. It is still being intensively developed and get many different variants of GAN. GAN was proposed to generate similar-looking samples to those in the training data sets. The emergence of GAN and its variants also provide new ideas for food pairing. In this paper, we have tried to invent a novel technique for food pairing using GAN and its variants. Specifically, we adopted the Long Short Term-Memory (LSTM) as the generator and the Convolutional Neural Network (CNN) as the discriminator. The sequences of recipes as the input will be encoded by LSTM into target sequences, which were finally identified by CNN to compute the differences between the generated recipes and their original input. The CNN will give a feedback to LSTM to optimize its parameters until the end of training process. As different customers have different food tastes, we have improved our method and invented new model using Conditional GAN (CGAN) to incorporate the personal demands in food pairing. We have conducted extensive experiments on real data sets to evaluate the efficiency of our proposed methods. The experimental results proved that our methods can generate better food pairings.

Keywords: Food pairing · Recipes · GAN · LSTM · CNN · CGAN

1 Introduction

Food is one of the most important parts in our daily life. There are many kinds of foods in China. From ancient times to today, different food styles and food cultures have been formed in different parts of China. The tastes of local cuisines are also very different. So, can we try to develop new recipes using new technologies based on existing recipes? In recent years, deep neural networks have made significant progress in many applications, including image generation, speech recognition, text generation and machine translation. The development of text generation techniques based on deep neural networks has important practical

H. Mei et al. (Eds.): BigData 2020, CCIS 1320, pp. 148–164, 2021.
https://doi.org/10.1007/978-981-16-0705-9_11

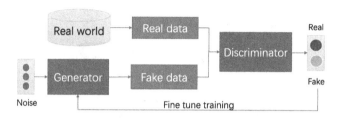

Fig. 1. GAN training framework

implications for developing new recipes, enriching recipe diversity, and reducing food wastes. We hope to use the deep learning methods to find the inner relationships among different ingredients from exiting recipes. Based on the new findings, we try to invent novel methods for developing new recipes.

Recently, Generative Adversarial Networks (GAN) [4] have received a great deal of attention from academic and industry. By designing a *minimax* game between a generative model and a discriminative model, GAN and its variants have achieved great successes in various applications [15], such as image generation [2], sequence generation [17], dialogue generation [11], information retrieval [16], and domain adaption [21]. In these applications, GAN has achieved better results in text generation and produced a series of improved models, such as TextGAN [19], SeqGAN [17], MaskGAN [3], and LeakGAN [6]. In fact, food pairing (recipe generation) is also a text generation task. So, we try to invent novel methods for new recipes generation using GAN and its variants.

GAN consists of a generator and a discriminator as shown in Fig. 1. The generator is used to capture the data distribution. The discriminator is used to classification or recognition. In fact, the discriminator can be a traditional machine learning algorithm, such as SVM, or deep learning algorithm, such as CNN [5]. As shown in Fig. 1, in the training process, the goal of the generator is to generate fake data as much as possible to deceive the discriminator. The goal of the discriminator is to distinguish the fake data generated by the generator from the real data. During training process, the discriminator must be synchronized well with the generator.

As an unsupervised learning model, GAN is characterized by the ability to learn from completely random raw data. However, this learning process is usually very inefficient. If we want to use GAN for text generation, there are many challenges to solve. For example, humans can dynamically measure the quality of text at any time, such as whether the semantics are complete, whether the grammar is correct, and whether the content has been clearly expressed. But, it is very difficult for artificial intelligence systems to do these works. In fact, food pairing is also a text-generating task. The referred problems are encountered when using GAN for food pairing. In order to address the disadvantages of GAN that cannot directly handle sequence tasks, we adopted SeqGAN.

SeqGAN adopts LSTM [8] as the generator because of LSTM can avoid the problem of gradient disappearance of RNN. And, SeqGAN adopt CNN-based

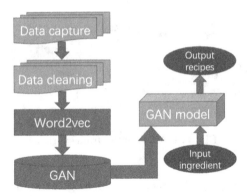

Fig. 2. Work flow of food pairing using GAN

binary classification model as the discriminator because it is superior to other algorithms in text classification [18]. The work flow of this work is shown in Fig. 2. After data capture and cleaning, we can get the training data set required for the experiment. Then, we can get a trained model for food pairing by using GAN to learn the inner relationships from the training data set. Finally, we can reuse the trained model to create innovative recipes based on a given ingredient that will be paired with some other ingredients recommended by the model.

2   Related Works

In recent years, deep generative models have drawn more and more attention because of its flexible learning ability for a large number of unlabeled data. In 2006, Geoffrey Hinton first proposed to use the contrastive divergence algorithm to train deep belief nets (DBN) efficiently [7]. For each data instance, DBN learns a low dimensional representation using encode network and generate new data instance using a decoding network. Variational auto-encoder (VAE) is another deep generative model, which combines deep learning with statistical inference and intends to represent a data instance in a latent hidden space using neural networks for non-linear mapping [9]. Both DBN and VAE are trained by maximizing training data likelihood, which suffers from the difficulty of approximating intractable probabilistic computations.

In 2014, Ian Goodfellow proposed GAN [4], an alternative training methodology to generative models. The training procedure of GAN is a *minimax* game between a generative model and a discriminative model. This framework avoids the difficulty of maximum likelihood learning and is one of the most promising methods for unsupervised learning in complex distribution in recent years.

However, GAN has problems with text generation. The image generation with good results of GAN belongs to continuous data generation, which can be optimized by direct parameters, and finally generate images with false realism. But the text data is discrete data, in which case the discriminator cannot propagate the gradient back to the generator. In [19], GAN was applied to text generation, and the TextGAN model was proposed. The idea of smooth approximation was used to approximate the output of the generator, thus solving the problem of gradient non-conductivity. The author's optimization function used the feature matching method and used a variety of training methods to improve the convergence of GAN.

The literature [20] is a continuation of the literature [19], which focuses on the process of confrontation training. The objective function during training is not the objective function of the original GAN, but a match operation is performed on the hidden feature representation of the real sentence and the generated sentence through the kernelized discrepancy metric. This approach can alleviate the problem of mode collapse in confrontation training.

In [17], the authors proposed SeqGAN. The starting point of this paper is the difficulty that GAN will encounter when dealing with discrete data. Mainly reflected in two aspects: the generator is difficult to pass gradient updates, while the discriminator is difficult to evaluate non-complete sequences. For the former, the author regards the whole GAN as a Reinforcement Learning (RL) system, and updates the parameters of the generator with the Policy Gradient algorithm [14]. For the latter, the author draws on the idea of Monte Carlo tree search (MCTS) to evaluate non-complete sequences at any time.

In [3], the authors proposed MaskGAN to accomplish tasks similar to cloze. The authors use the actor-critic algorithm in RL to train the generator. The discriminator is trained by using maximum likelihood and stochastic gradient descent. GAN's mode collapse and training instability problems are also serious in text generation tasks. To avoid the impact of these two problems, the authors fill in missing words based on the context.

In [6], the authors proposed LeakGAN to solve the problems of long text generation. The discriminator leaks some extracted features to the generator at the intermediate time step, and the generator uses this additional information to guide the generation of the sequence. The generator uses a hierarchical reinforcement learning structure, including the Manager module and the Worker module. The Manager module is based on LSTM. At each time step, the Manager module receives a feature representation from the discriminator and then passes the feature representation as a guidance signal to the Worker module. The Worker module then uses LSTM to encode the input, and the output of LSTM is connected to the received guidance signal to select next word.

Based on the analysis of the existing works, this paper uses the SeqGAN to study the food pairing. The generator uses an LSTM network-based model, the sequences of ingredients as the input, and a variety of ingredients that can be paired with input as the output. The discriminator is based on the CNN's two-category model, which calculates the similarity between the recipes generated by the generator and the real recipes. In order to satisfy the customer's personal demands on food tastes, we improved our model using CGAN.

Fig. 3. SeqGAN model

3   Food Pairing Using SeqGAN

As shown in Fig. 1, the generator and the discriminator of GAN trained together and finally get a trained generation network. The trained network can produce fake data that similar to the real data in distribution with high probability. In the learning process, the generator will generate fake data based on the real data. The generated fake data combined with the real data will be input to the discriminator. Next, the discriminator will identify the authenticity of the input data. The evaluated results will be sent to the generator as the feedback. Then, the generator will optimize its parameters based on the feedback and generate new fake data. The training process will be continued until the discriminator cannot identify the authenticity of the input data.

However, traditional GAN often has problems such as gradient disappearance and mode collapse during training process. These problems become very critical when processing discrete data, such as text. In order to solve the referred problems, SeqGAN [17] was proposed. Regarding to the advantages of SeqGAN, we apply it for food pairing task in this work. The framework of SeqGAN is shown in Fig. 3. The architectures of its generator and discriminator are given in Fig. 4 and Fig. 5, respectively.

3.1   SeqGAN Model

The most straightforward way to build SeqGAN model is to use Recurrent Neural Network (RNN) as the generator, because RNN has achieved good results in multiple tasks of natural language processing. However, RNN has many disadvantages, such as gradient disappearance and gradient explosion during long sequence training. Long Short-Term Memory (LSTM) as one of the variants of RNN can avoid its problems. So, selecting LSTM as the generator of SeqGAN is a better way. But, there is a challenge should be resolved. LSTM cannot adopt the discriminator's feedback to tune its gradient and to guide the generation of discrete data. Inspired by the ideas of Reinforcement Learning (RL), SeqGAN consider the output of the generator as a sequence of decisions on character selection tasks. At the same time, they consider the probability estimation by the discriminator of the generated sample as the feedback signal for the generator. In this way, SeqGAN can use the parameter adjustment technique of RL to tune the parameters of the generator for text generation.

When choosing discriminator, SeqGAN uses CNN as its discriminator. CNN applies several numbers of kernels with different window sizes to extract features and applies a max-pooling operation over the feature maps. And, it selects ReLu as its activation function. As a completely unsupervised training process, a large number of experiments show that ReLu's output is closer to or even beyond the pre-training effect, and has a faster training speed [10]. In order to make the classification of the model better, there is a highway network to CNN, as shown in Fig. 5. Finally, a fully connected layer with sigmoid activation is used to output the probability that the input data is real.

3.2 SeqGAN Pre-training

In order to improve the efficiency of the generator training, there is a pre-training process in SeqGAN framework.

We use the maximum likelihood estimation (MLE) to pre-train the generator. Then, we pre-train the discriminator via minimizing the cross entropy.

In order to guarantee the accuracy of the discriminator, we need the same number of positive samples and negative samples. The negative samples are generated by the generator.

3.3 SeqGan Training

After the pre-training process, generator and discriminator are trained alternatively. In this process, both parties strive to optimize their network: the generator continuously optimizes its ability to generate realistic text, and the discriminator continuously improves the ability to distinguish between the generated text and the real text, thus forming a competitive confrontation until the two sides reach the Nash equilibrium.

(1) The Generator of SeqGAN

During the training process of the generator, each of ingredients in the training data is mapped to a vector by embedding layer, and then generate an encoding matrix. This representations help SeqGAN to learn the meanings of each ingredient and produce reasonable results. As shown in Fig. 4, we input the recipes sequences to generator. Then, we obtain the embedding of each ingredient and input the embedding into each cell.

LSTM can delete or add information to the cell state through a structure called "gate" [8]. LSTM includes three gate structures, namely forget gate, input gate and output gate. The forget gate decides to discard some information. The input gate determines which information is updated. And the output information is determined by the output gate. Finally, the output information is determined by the output gate, and then the vector representation of the output information is obtained. Combined with a fully connected layer, we could generate fake recipes.

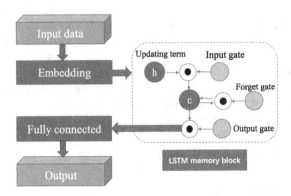

Fig. 4. The generator of SeqGAN

(2) The Discriminator of SeqGAN

During the training process of the discriminator, the input data is a two-category recipes. As shown in Fig. 5, one is real recipes, the other is generated recipes by the generator. The training process of the discriminator is supervised training. We use the real recipes as positive samples and set the label to 1. In the same way, we use the generated recipes as negative samples and set the label to 0. We use the cross entropy (between the ground truth label and the predicted probability) loss function to calculate the loss value of the discriminator.

The specific training process of the discriminator is as follows. We convert the vector of the input sequence into matrix. The discriminator applies a convolution kernel with a window size of l words. The convolution kernel obtains the feature map by convolution with the input recipe matrix. Then, we apply pooling operations to feature map. The discriminator outputs a scalar that represents the probability that the input data is a real recipe. After training, the discriminator can distinguish the real recipes from the generated recipes with the greatest precision.

3.4 Evaluation Methods

The model effect evaluation of generative models has always been an unsolved problem. There is no uniform evaluation index in the field of image generation or text generation. It is difficult to quantitatively describe the similarity between generated data and real data, and this quantitative description needs to be consistent with human views. In this work, we adopted two evaluation methods, one is based on similarity computation and another is based on manual evaluation.

(1) Similarity Computation

In order to evaluate our model, we use Bilingual Evaluation Understudy (BLEU) [13] score as an evaluation metric to measure the similarity degree between the generated recipes and the real recipes. The greater the similarity degree between the two data sets, the better the generator results.

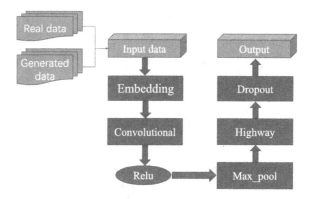

Fig. 5. The discriminator of SeqGAN

In the machine translation model, BLEU is used to evaluate the quality of translated text. That is, the similarity between the text translated by the machine and the text translated by the human experts. Although this assumption ignores the diversity of reasonable translation, BLEU is still one of the best evaluation methods in machine translation in the case of enough reference translations.

The output of BLEU is always a number between 0 and 1, which represents the similarity between the generated text and the reference text. The output value of BLEU is 1 when the generated text is exactly the same as one of the samples in the reference text set. In subsequent experiments, this article will use the "Score" to refer to the evaluation indicators introduced in this section.

(2) Manual Evaluation

In order to ensure that the evaluation results are more reasonable, we also use manual evaluation methods. We set up three evaluation indicators for the experiments, which are recipe integrity, without duplicate ingredients and edible. The score for each indicator is between 0 and 1. The higher the score, the better the recipe generated. The meanings of each indicator are described as following.

1. **Recipe Integrity.** Each recipe contains basic major ingredients and minor ingredients.
2. **Without Duplicate Ingredients.** The same ingredients will not be repeated in each generated recipe.
3. **Edible.** Foods that appear in the same recipe can be paired with each other.

We have three professionals to evaluate the generated recipes and set scores to the three indicators above. Among the three indicators, recipe integrity and edible are set according to the recipes used in this experiment. For the indicator 1, most recipes should contain several basic ingredients, such as *salt*, *MSG*, *onion*, *ginger*, *garlic*, and *cooking oil*. The more complete of the ingredients contained in the recipe, the closer to 1 of the indicator 1. In addition, the indicator 2 is

Fig. 6. CGAN model

used to reflect the training efficiency of the model. In general, we do not hope the same or similar ingredients to appear repeatedly in the same recipe. The indicator 3 measures the reasonableness of the food pairing. According to the real recipes, the more reasonable of the recipe, the higher of the score of the indicator 3.

4   Food Pairing with Personalized Requirement

People from different regions has different tastes. For example, people in *Shanxi* prefer *sour* taste, people in *Sichuan* and *Chongqing* prefer *spicy* taste, and *Jiangsu* and *Zhejiang* prefer *sweet* taste. Moreover, everyone has their own preferences or needs for food. So, we should provide different food pairing and cooking methods. Given one or several major ingredients and tastes, we can provide multiple recommendations, including the minor ingredients and other necessary major ingredients.

As the different minor ingredients can bring different tastes, the same group of major ingredients are usually cooked with different minor ingredients to get different tastes. For example, when *egg* is cooked with *rice vinegar, chili sauce*, and *shallot*, the taste is *spicy* and *sour*. When *egg* is cooked with *salt, shallot*, and *spiced salt*, the taste is *salty*. When *egg* is cooked with *sugar* and *tomato sauce*, the taste is *sweet* and *sour*.

As the common GAN model cannot incorporate the input constraints, we try to apply the CGAN model to provide the food pairing recommendations with personalized requirement. The architecture of CGAN is shown in Fig. 6. Also, we formalized the training recipe data to fulfill the requirements of CGAN.

4.1   CGan Model

In order to meet the needs of all users, we need to put some constraints on GAN. In this section, we applied Conditional Generative Adversarial Networks (CGAN) [12]. CGAN can incorporate conditional variables in its generator and discriminator. Using these conditional variables to add constraints to the model can guide the data generation process. If the input constraint variable is category label, we can consider CGAN to be an improvement in turning GAN into a supervised model.

So we draw on the ideas of CGAN to design and train a new network model for food pairing to incorporate the personalized requirement. By doing this, the generated recipes become more practical than using GAN.

Similar to GAN, CGAN retains the game structure of generator and discriminator. The difference is that CGAN adds conditional input to both the generator and discriminator inputs. On the input side of the generator, the noise is combined with the constraints as the input of the generator, and the fake samples are generated by the generator. The discriminator not only needs to identify the similarity between the generated fake sample and the real sample, but also needs to identify whether the generated sample meets the constraints.

The generator of CGAN uses LSTM. The essence of the discriminator is to extract and distinguish features from real samples and generated samples. Therefore, the discriminator of CGAN still uses CNN.

4.2   Food Pairing Using CGAN

We trained the conditional generative adversarial network on recipes paired with their taste labels as the input constraints.

In the generator of CGAN, both noise and their labels are mapped as the input of the hidden layers of LSTM. Through training, we have a generator that could generate realistic enough recipes under given condition.

Similarly, in the discriminator of CGAN, the recipes and its taste labels will also be connected together as the input of the hidden layer of CNN. Through training, the discriminator can well judge whether the sample is real recipe that meets the condition.

Then we save the trained CGAN model locally. By calling this model, we can input a desired taste and ingredient, then output other ingredients that meet the taste and can be pairing with the input ingredient.

5   Experiments

In this section, we have designed four experiments to evaluate the food pairing efficiency of our proposed methods. (1) The effects of hyperparameter k on loss changes. (2) The evaluation of food pairing results generated using SeqGAN. (3) Comparisons with other deep model. (4) The evaluation of food pairing results generated using CGAN.

The experiments are conducted on Ubuntu 16.04 system, equipped with 40 cores, 256G main memory and Titan V GPU. The softwares installed are CUDA, CUDNN and TensorFlow [1].

The data set used in this work is scrapped from the China Food Formulation website[1]. The site offers a number of services, including recipe searches, Chinese cooking methods and recipe recommendations. After data cleaning, we obtained 10,000 recipes belonging to 20 different Chinese cuisines. Each recipe contains

---

[1] www.meishij.net.

multiple attributes such as "Ingredients", "Cooking methods", "Cooking times", "Taste" and "Tags". In our research, we use "Ingredients" and "Taste" to suit the computational requirements of the proposed algorithm. The attribute "Ingredients" represents a range of ingredients contained in each recipe, in which the ingredients are divided into major ingredients and minor ingredients.

Based on our experimental experiences, we set the related parameters to their optimal values at which the models can get the best efficiency. The dimension of the memory cell of the LSTM is set to 32. The gradient clip in back propagation is set to 5.0. The learning rate is set to 0.01.

5.1   The Effects of Hyper-parameter k

GAN's generator and discriminator share a set of parameters. If the training between the generator and the discriminator is not balanced, the discriminator will prematurely lose the ability to identify the real sample and the generated sample. The generator also stops optimization due to the discriminator stopping optimization, and eventually converges to equilibrium in a premature manner, resulting in poor quality of the generated samples.

In the training process of GAN, we trained the generator and the discriminator alternately. We use k to represent the training times of the discriminator in each alternative training process. That is, we alternate between k-step of optimizing discriminator and 1-step of optimizing generator [4]. As long as the generator changes slowly enough, the result of the discriminator can maintain its most recent optimal solution. The selection of the hyper-parameter k affects the time it takes for the parameter to converge, and the effect of the generator generating the recipe after the parameter tends to balance.

The specific training rounds ratio varies with the network structure and the complexity of the corpus, and needs manual adjustment. The specific training process can be split into two steps as following.

Step A: Sampling from the experimental data set to get the real recipes, and then sampling the generated recipes from the generator. The discriminator parameters are updated by supervising the discriminator.
Step B: Sampling the generated recipes from the generator, updating the parameters of the generator to increase the rewards of the newly generated recipes output in the discriminator.

As mentioned above, the discriminator is often difficult to train, so the k value should be increased. We analyzed the loss trend of the generator at different k values. The values of k are set to 1, 3, 5, and 8, respectively. Figure 7 shows the loss trends of SeqGAN at different k values.

In practice, we find that the discriminator is always weaker than the generator in the case of 1:1 update ratio, at which the network converges prematurely and cannot be fully trained. When $k = 3$, the model training is unstable. As shown in Fig. 7, when $k = 8$, the model converges quickly, but the quality of the recipe is not good enough. For example, when input *egg*, the BLEU score of the generated

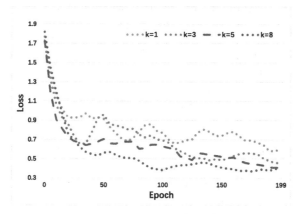

Fig. 7. Trends of loss at five k values

Table 1. Generated recipes for *Egg*

K	Minor ingredients	Major ingredients	Score
5	salt, cooking oil, cooking wine, pepper, sugar, sesame oil	—	0.86
	salt, blending oil	green peppers, tomato	0.80
	salt, onion, ginger, cooking wine, blending oil	pepper, noodles	0.75
8	salt, blending oil, onion	carrot, bell pepper	0.83
	salt, blending oil	shrimp, potato diced	0.80
	salt, blending oil, garlic, yellow wine, five-spice powder	bell pepper	0.62

recipes is low as 0.62, as shown in Table 1. By repeated attempts, k is finally set to 1:5, at which the discriminator is trained 5 times when the generator is trained once.

5.2   The Evaluation of Food Pairing Using SeqGAN

Based on the experimental observations, we trained SeqGAN with the training round ratio at 1:5 for the generator and the discriminator.

In order to verify the food pairing efficiency of SeqGAN, we selected three common major ingredients as the input to the model for the test. For each input ingredient, the trained model generates three different recipes. The experimental results are shown in Table 2.

By analysing the experimental results, it can be clearly observe that trained SeqGAN model can generate the recipes with a variety of common major ingredients and minor ingredients with high BLEU scores, as shown in Table 2.

Table 2. Food pairing using SeqGAN

Input ingredient	Minor ingredients	Major ingredients	Score
green pepper	salt, garlic, soy sauce, white sugar	cabbage	0.83
	salt, chicken essence, onion, ginger, garlic, starch	egg	0.75
	salt, blending oil, onion, ginger	carrot, cucumber	0.71
egg	salt, cooking oil, cooking wine, pepper, sugar, sesame oil	—	0.86
	salt, blending oil	green pepper, tomato	0.80
	salt, onion, ginger, cooking wine, blending oil	pepper, noodles	0.75
carrot	salt, soy sauce, onion, ginger, blending oil	green pepper	0.88
	cardamom	bell pepper, potato, pork belly, vermicelli	0.83
	salad oil, ginger, garlic, cooking wine, blending oil	cucumber, kumquat	0.75

The model can accurately learn the structure of many recipes and can pair ingredients that often appear on the same dish to create new recipes.

Moreover, we can draw the following conclusions from the experimental results.

1. *Salt, onion, blending oil* and *ginger* are used as minor ingredients in almost recipes.
2. *Blending oil* is used more frequently than other edible oils.
3. *Egg* and *green pepper* often appear together. Moreover, *carrot* and *green pepper* often appear together.

The above conclusions are also consistent with our daily cooking habits. So it can be considered that the generation effect of our model is very good.

5.3 Comparisons with Other Deep Model

In this experiment, we compare SeqGAN with RNN that is widely used in text generation task. The RNN is based on the TextgenRNN model, that is built on Keras and TensorFlow and can be used to generate character-level and text-level text. And, it uses attention weighting to speed up the training process and improve text quality. In the experiment, we use the different two evaluation methods: manual evaluation and BLEU algorithm. The results are shown in

Table 3. Comparisons with SeqGAN and RNN

SeqGAN generated recipes	BLEU	Indicator	RNN generated recipes	BLEU	Indicator
salt, garlic, soy sauce, white sugar, **green pepper**, cabbage	0.83	0.85	salt, oyster sauce, soy sauce, chicken essence, starch, ginger, garlic, potatoes, eggplant, **green pepper**	0.79	0.90
salt, chicken essence, onion, ginger, garlic, starch, **green pepper**, egg	0.75	0.80	salt, cooking wine, soy sauce, oyster sauce, chicken essence, pepper, sesame oil, pork belly, **green pepper**	0.46	0.80
salt, cooking oil, cooking wine, pepper, sugar, sesame oil, **egg**	0.86	0.80	salt, soy sauce, MSG, ginger, sesame oil, white sugar, dried mushrooms, broad bean starch, **egg**, yellow wine	0.81	0.85
salt, blending oil, **egg**, green pepper, tomato	0.80	0.60	chicken essence, blending oil, yuan sugar, milk, **egg**	0.80	0.50
salt, soy sauce, onion, ginger, blending oil, **carrot**, green pepper	0.88	0.90	salt, soy sauce, sesame oil, MSG, onion, ginger, cooking wine, lemon, sugar, egg, pork, **carrot**	0.75	0.80
carrot, cardamom, bell pepper, potato, pork belly, vermicelli	0.83	0.50	blending oil, yuan sugar, yam, **carrot**	0.75	0.40

Table 3. In the table, the indicator is the average of the three indicator scores. The ingredients in bold are the input for comparison.

From Table 3, we can find that in these recipes, the BLEU score of SeqGAN is stable around 0.8, and most of the scores are higher than 0.8. Similarly, the BLEU score of RNN is stable between 0.75 and 0.8, and the BLEU score of one recipe is as low as 0.46. We can think that the BLEU score of SeqGAN is almost higher than or equal to the BLEU score of RNN. For Indicator Score, the performance of these two models is similar. In general, the performance of the SeqGAN is better than RNN. The generated recipes by SeqGAN are relatively complete and reasonable.

Table 4. Food pairing using CGAN

Input ingredient	Input taste	Minor ingredients	Major ingredients	Score
shiitake	salty	salt, MSG, ginger, cooking Wine, starch, pepper	peanuts, celery	0.80
	spicy	salt, soy sauce, MSG, shallot, ginger, starch, chili oil, lard, sesame oil	chicken	0.75
	homely flavor	salt, soy sauce, MSG, ginger, sugar	bell pepper, ham, bamboo shoot, pork chop	0.73
eggplant	salty	salt, soy sauce, onion, garlic, rice wine	egg	0.87
	garlic fragrance	salt, onion, ginger, garlic, cooking wine, sugar	—	0.87
	homely flavor	salt, chicken essence, light soy sauce	tricholoma	0.67
carrot	salty	salt,onion, ginger	scallop, northern tofu	0.85
	spicy	salt, onion, ginger, soy sauce, MSG, sesame oil, sugar, chili powder	egg, bell pepper	0.81
	homely flavor	salt, vegetable oil	beef, quail egg	0.83

5.4   The Evaluation of Food Pairing with Personal Requirement

In this experiment, we evaluate the efficiency of CGAN in food pairing with personal requirement. We select shiitake, eggplant and carrot as input ingredients, separately. For each input ingredient, we set three different tastes as personal requirement, as shown in Table 4.

From the Table 4, we can see that CGAN performs well in most cases with high BLEU scores. And, the generated recipes for each input with different tastes are complete, reasonable and edible. When the input taste requirement is *spicy*, the recommended minor ingredients always contain chili oil or chili powder. When the input taste requirement is *garlic fragrance*, the recommended minor ingredients always contain *garlic*.

Moreover, we can see that some minor ingredients such as *salt, MSG, onion, ginger, soy sauce*, etc. are present in most output recipes.

6   Conclusions and Future Work

Text generation using GAN is a relatively new research direction and has a broad applications. Unlike traditional food pairing methods, using GAN and its variants for food pairing is novel. As the recipes are usually invented by experienced cookers and exist hundreds of years. In the existing recipes, there are many hidden patterns, including culture, religion, geography, climate and so on. GAN and its variants can learn inner relationships between real recipes and produces new recipes. In order to take the personal requirement into food pairing, we have tried the CGAN model. We evaluated our proposed methods by using the BLEU algorithm and our proposed manual evaluation method. We conducted extensive experiments on real data set. The experimental results shown that our proposed novel food pairing methods can generate complete, reasonable and edible recipes.

As well known, the challenging problems of GAN and its variants are training instability and pattern collapse. In the future work, we will try to achieve stable training and obtain a well-diversified model. Also, we will try to speed up the training process.

Acknowledgment. This work was supported by the project of Natural Science Foundation of China (No. 61402329, No. 61972456) and the Natural Science Foundation of Tianjin (No. 19JCYBJC15400).

References

1. Abadi, M., et al.: TensorFlow: a system for large-scale machine learning. In: OSDI 2016, pp. 265–283 (2016)
2. Denton, E.L., Chintala, S., Szlam, A., Fergus, R.: Deep generative image models using a Laplacian pyramid of adversarial networks. In: NIPS 2015, pp. 1486–1494 (2015)
3. Fedus, W., Goodfellow, I.J., Dai, A.M.: MaskGAN: better text generation via filling in the _. In: ICLR 2018 (2018)
4. Goodfellow, I.J., et al.: Generative adversarial nets. In: NIPS 2014, pp. 2672–2680 (2014)
5. Gu, J., et al.: Recent advances in convolutional neural networks. Pattern Recog. **2018**(77), 354–377 (2018)
6. Guo, J., Lu, S., Cai, H., Zhang, W., Yu, Y., Wang, J.: Long text generation via adversarial training with leaked information. In: AAAI 2018, pp. 5141–5148 (2018)

7. Hinton, G.E., Osindero, S., Teh, Y.W.: A fast learning algorithm for deep belief nets. Neural Comput. **2006**(18), 1527–1554 (2006)
8. Hochreiter, S., Schmidhuber, J.: Long short-term memory. Neural Comput. **1997**(9), 1735–1780 (1997)
9. Kingma, D.P., Welling, M.: Auto-encoding variational Bayes. In: ICLR 2014 (2014)
10. Krizhevsky, A., Sutskever, I., Hinton, G.E.: ImageNet classification with deep convolutional neural networks. Commun. ACM **60**, 84–90 (2017)
11. Li, J., Monroe, W., Shi, T., Jean, S., Ritter, A., Jurafsky, D.: Adversarial learning for neural dialogue generation. In: EMNLP 2017, pp. 2157–2169 (2017)
12. Mirza, M., Osindero, S.: Conditional generative adversarial nets. CoRR 2014 abs/1411.1784 (2014)
13. Papineni, K., Roukos, S., Ward, T., Zhu, W.J.: BLEU: a method for automatic evaluation of machine translation. In: ACL 2002, pp. 311–318 (2002)
14. Sutton, R.S., McAllester, D.A., Singh, S.P., Mansour, Y.: Policy gradient methods for reinforcement learning with function approximation. In: NIPS 1999, pp. 1057–1063 (1999)
15. Wang, H., et al.: GraphGAN: graph representation learning with generative adversarial nets. In: AAAI 2108, pp. 2508–2515 (2018)
16. Wang, J., et al.: IRGAN: a minimax game for unifying generative and discriminative information retrieval models. In: SIGIR 2017, pp. 515–524 (2017)
17. Yu, L., Zhang, W., Wang, J., Yu, Y.: SeqGAN: sequence generative adversarial nets with policy gradient. In: AAAI 2107, pp. 2852–2858 (2017)
18. Zhang, X., LeCun, Y.: Text understanding from scratch. CoRR 2015 abs/1502.01710 (2015)
19. Zhang, Y., Gan, Z., Carin, L.: Generating text via adversarial training. In: NIPS 2016, vol. 21 (2016)
20. Zhang, Y., et al.: Adversarial feature matching for text generation. In: ICML 2017, pp. 4006–4015 (2017)
21. Zhang, Y., Barzilay, R., Jaakkola, T.S.: Aspect-augmented adversarial networks for domain adaptation. TACL **5**, 515–528 (2017)

Comparisons of Deep Neural Networks in Multi-label Classification for Chinese Recipes

Zhaopei Liu[2], Chuitian Rong[1,2(✉)] [iD], and Xiangling Zhang[3]

[1] Tianjin Key Laboratory of Autonomous Intelligence Technology and Systems, Tianjin, China
[2] School of Computer Science and Technology, Tiangong University, Tianjin, China
`chuitian@tiangong.edu.cn`
[3] Beijing Institute of Education, Beijing, China
`zhangxiangling@bnu.edu.cn`

Abstract. With the highly increasing demands, multi-label classification task has attracted more attention in recent years. However, most of traditional methods usually need tedious handcrafted features. Motivated by the remarkably strong performance of deep neural networks on the practically important tasks of natural language processing, we adopted various popular models, such as CNN, RNN and RCNN, to perform multi-label classification tasks for Chinese recipes. Based on the real Chinese recipe data extracted from websites, we compared the performance of deep neural networks in multi-label classification. We also compared them with the baseline models, such as Naive Bayes, MLKNN and fastText. In order to improve the performance of the these models, we adopted the data augmentation method and then conducted extensive experiments to compare different models in our task. The results showed that RCNN model performs the best and can get the highest score. The models based on deep neural networks all performed better than the baseline models. The results also showed that the data augmentation method is a practical method to improve the performance of all models.

Keywords: Multi-label classification · Deep neural networks · Data augmentation

1 Introduction

Nowadays, multi-label classification has aroused widely attention on many social networks and applications, like LinkedIn and Mircoblog. An important reason is that the usage of multi-label classification on them not only help users rapidly find their favor contents, but also help the website developer to integrate similar resources and provide personalized content services. Another reason is that only a few labels of microblogs or other contents are created by their authors. How

© Springer Nature Singapore Pte Ltd. 2021
H. Mei et al. (Eds.): BigData 2020, CCIS 1320, pp. 165–179, 2021.
https://doi.org/10.1007/978-981-16-0705-9_12

to perform multi-label classification automatically for the contents online is getting more and more important. Labels can appear everywhere, which represent specific features of the related contents. For the Chinese recipes data we collected, the labels behind them refer to the dietary functions, such as "Calcium supplement" label in "Mapo Tofu".

Previous studies for multi-label classification have proposed various traditional methods, such as KNN, SVM and Logistics. Most of these methods need the complex and tedious feature engineering. They commonly need to transform data into one-hot representation, which will cause semantic loss. Meanwhile, when deal with multi-label classification tasks, most of the traditional methods need to be transformed into multiple binary classifiers. However, there are some connections between labels, the weakness of traditional multi-label method is that they ignore the relationship between labels.

In the last few years, deep neural networks with word embedding has achieved high performance in various NLP tasks. As a typical representative, convolutional neural networks (CNNs), which was originally proposed for computer vision, have shown their effectiveness for various NLP tasks, including semantic analysis, machine translation, topic classification, and a variety of traditional NLP tasks. Instead of building hand-crafted features, these methods utilize layers with convolving filters that are applied on top of pre-trained word embeddings.

Inspired by the remarkable performance of deep neural networks on NLP tasks, we conduct our research by using CNNs, RNNs, RCNNs with word embedding to solve with the multi-label classification task for Chinese recipes. Moreover, previous studies have demonstrated that data augmentation can improves performance for deep neural networks, particularly to strengthen the performance for smaller datasets. So, we also adopted data augmentation in our experiments. As the baselines of our experiments, we compared the deep neural models with the traditional models, including MLKNN, Naive Bayes and fastText. In our work, we first use pad-sequence method to ensure the same length of each recipe content. Second, we use the word2vec to train the processed ingredients words and get an extended ingredient sentiment dictionary. Third, we use 80% of our recipe datasets as the training data, and use 10% of the training data as the validation sets to adjust learning rate in the training process. Besides, we use the remaining 20% recipe datasets as the test data set to estimate the effectiveness and accuracy of these deep neural networks on this task. Finally, we adopted data argumentation in data preparation process and repeated the steps above to compare the performance of these methods on multi-label classification for Chinese recipes.

The main contributions of our study can be summarized as follows:

- To take advantages of deep neural networks, we adopted CNNs, RNNs and RCNNs to perform this multi-label classification for Chinese recipes.
- For better performance, we adopted data augmentation method in data preparation process and then we conducted extensive experiments in all proposed models. The results demonstrated that deep neural networks can achieve better performance after using data augmentation.

– We also compared the performance of deep neural networks with the baseline models. The result showed that models based on deep neural networks can perform better than traditional machine learning methods.

2   Related Works

2.1   Multi-label Classification

Multi-label classification is an important part of NLP. For some texts with Multi-labels, it allows several labels to be assigned to one sample text. The relationship between labels is usually hierarchical, which can be used to describe the attributes of the text from multiple dimensions. The implementation of multi-label classification algorithm can be divided into supervised learning and unsupervised learning. At the same time, there are summarily two traditional ways to deal with multi-label classification task. One is to transform multi-label classification task into multiple binary classification tasks, then form a classification chain to predict the final classification results. The second way is directly changing the algorithm in order to adapt the multi-label classification task.

For the first way, Read, J. et al. proposed multi-label classification chains [13]. These chains can overcome the shortcomings of the binary association method by passing label related information along the classifier chain, which obtains higher prediction performance with the lower computational complexity. Unlike a specialized adaptive classification algorithm, the classifier chain can be seen as a off-the-shelf method that does not require a parameter configuration, and has robust performance in a range of data. For high performance, they also have high scalability that can be used on the largest set of data we consider. G. Tsoumakas et al. proposed a RAKEL method [16] that learns a set of LP classifiers, each of which corresponds to a different small random subset of a set of labels. Meanwhile, the method also takes into account the calculation efficiency and the prediction performance of the standard LP method in the case of a large number of labels and training examples.

For the second way, K. Dembczynski et al. proposed a Bayesian optimal multi-label classification [2] which is based on probability classifier chain. Their method first estimate the whole joint distribution of labels, then can adapt different loss functions, and finally balance the complexity and the accuracy of probability estimation in an optimal way. They also studied the dependence between labels in multi-label classification [3]. And Ji, S. et al. proposed a framework [4] for extracting shared structures from multi-label classification. In their framework, the information of multiple labels is captured by the low dimensional subspace, which shared in all labels. Their research shows that when the least square loss function is used in the classification, the shared structure can be calculated by the generalized eigenvalues. Xia, X. et al. proposed RW.KNN algorithm [19] for multi-label classification. RW.KNN combines the advantages of KNN and Radom Walk algorithm, which provides a novel perspective in this research area. At the same time, they proposed a novel algorithm based on minimizing Hamming Loss to select the classification threshold.

Labels in multi-label datasets are always multi-hierarchical, when deal with the task of multi-label classification, the disadvantage of the two ways above is that they ignore the relationship between multiple labels, which will cause bias for the results of prediction. The deep learning method with word embedding as input shows excellent performance in the task of multi-label classification. Some methods arouse widely attention in NLP, including fastText, convolution neural network (CNN), recurrent neural network (RNN) and convolution recurrent neural network (RCNN). fastText [5], the method that learn sentence representations dietary, and it reduce the gap in accuracy between linear and deep models by incorporating additional statics like using bag of n-grams. Yoon Kim used CNN with word embedding [6] on a series of experiments for sentence classification [7], which shown excellent performance despite little tuning of hyperparameters. Pengfei Liu introduced three RNN based architectures to model text classification with multi-task learning [12]. Their experiments indicated that the joint learning of multiple related together can improve the performance of a group of related tasks by exploring common features. Siwei Lai et al. proposed RCNN for text classification [11], which captures contextual information with the recurrent structure and constructs the representation of text using a convolution neural network. Their experiments showed the excellent performance of RCNN.

2.2   Data Augmentation

Deep neural networks have shown higher performance in nature language processing. But high performance of them in a specific task is always dependent on high quality and large-scale training datasets, which wastes more times to prepare. Automatic data augmentation has appeared for recent years, which commonly be used in computer vision [10,15] and speech [1,8]. Automatic data augmentation has shown that it can help training more better models, especially when using smaller datasets. In contrast, because creating universal rules for languages transformation is pretty hard, generalized data augmentation in NLP has not be widely used.

There are some previous work about data augmentation in NLP. Kobayashi has proposed a novel data augmentation using predictive language model for synonym replacement [9], which improves classifiers using convolutional or recurrent neural networks in various classification tasks. Another work has considered noising primitives as a form of data augmentation for recurrent neural network-based language models. Other popular research has generated new data by translating the original sentences from English to another language and back into English. These techniques have shown their effectiveness on improving the performance of neural networks, but they have a high cost of implementation. That is the reason why they are not often be used.

Easy data augmentation [17] has shown remarkable performance on text classification tasks, which consists four simple but powerful operations. Hence, we adopt two operations of easy data augmentation in the data process of our work, including random swap and random deletion.

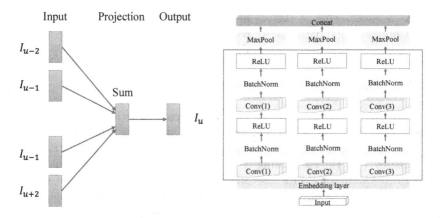

Fig. 1. CBOW model   **Fig. 2.** CNN model

3   Models for Multi-label Classification

In this section, we describe the models we adopted in multi-label classification for Chinese recipe, including word vector model, CNNs model, RNNs model and RCNNs model. We first train word vector model using all the ingredients of Chinese recipes. Then each recipe is transformed into vectors based on the word vector model. Besides, labels of recipes are also utilized to train the designed neural networks and then the labels of each test recipe data will be predicted according to the networks. Detailed information are demonstrated as follow.

3.1   Word Vector Model

The word2vec [14] tool named gensim is utilized to train the word vector, which provides CBOW and Skip-gram models. In our work, each ingredient of Chinese recipes is transformed into vector according to CBOW model. The CBOW model is proposed in Fig. 1, which can predict the current word based on the context. For a specific recipe $R = (I_1, I_2, ..., I_n)$, each ingredient I_i is represented as $I_i = (i_1, i_2, ..., i_k)^T$, where k is the dimension setted up in word vector model and i_i is the value in the ith dimension. The ingredients represented as vectors in the recipes are aggregated in order to keep the semantic features as follow,

$$R = I_1 \oplus I_2 \oplus ... \oplus I_N$$

where $\oplus$ is the concatenation operator. So a recipe R is transformed into a string of vectors as the order of ingredients. And for an document D, it is represented as follow,

$$D = R_1 \oplus R_2 \oplus ... \oplus R_K$$

where R_i is each recipe of the recipe space D. By the word vector model trained by Chinese recipe datasets, an recipe space is represented as a series of recipe, which can be used in deep neural networks.

3.2   CNN Model

After counting the recipe ingredient length, we observed that the major ingredient average length was 2.47, and the minor ingredient average length was 5.39. In that, the recipe ingredient data was a typical short text. So we extracted the local features after the embedding layer by uses multiple filters, varying the window sizes h. Comparing with the original TextCNN, we have made some improvements for this model, which is mainly manifested in: expanding the convolution layer from the single layer to the double layer, using BatchNorm instead of Dropout for regularization between the convolution layer and the activation function with ReLU, using double full connection layer instead of the single layer. Then, the features form the penultimate layer and act as the input of a fully connected softmax layer whose output is the probability distribution over all the labels. The CNN framework represented in Fig. 2.

In the details of CNN model, N-gram features are extracted by filters in the convolutional layer firstly. A convolution operation involves a filter $w \in \mathbb{R}^{hk}$, which is applied to a window of h words of k dimension to produce a new feature. In general, let $W_{i:i+h}$ refer to the concatenation of ingredients $I_i, I_{i+1}, ..., I_{i+h}$. So a feature c_i generated from a window of words $W_{i:i+h}$ can be represented as

$$c_i = f(w \cdot W_{i:i+h} + b)$$

where $b \in \mathbb{R}$ is a bias parameter and f is non-linear function such as sigmoid function and so on. This filter is applied to each window of ingredients of the recipes to produce map as

$$c = (c_1, c_2, ..., c_{N-h+1})$$

where $c \in \mathbb{R}^{N-h+1}$. In this work, we use the function of $ReLU(x) = \max(0, x)$. Then we repeat the process using filters with different sizes to increase the N-gram coverage of the model. The maximum pooling function is widely used. Specifically, it acts on the scores vector of the filter w to obtain the pooling score p_f as

$$p_f = \max(c) = \max(c_1, c_2, ..., c_{n-w+1})$$

3.3   RNN Model

Recurrent neural networks (RNN) are a type of network which form memory through recurrent connections. Comparing with the original TextRNN model, we performed k-MaxPooling on all hidden units of the model instead of directly using the last hidden unit as a classification. In feed forward networks, inputs

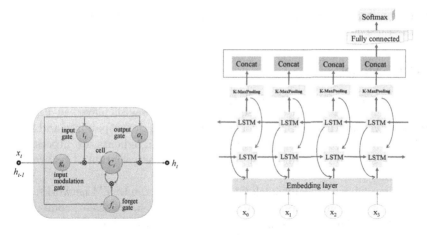

Fig. 3. LSTM cell   **Fig. 4.** RNN model

are independent of each other. But in RNN, all inputs are connected to each other. This enables the network to exhibit robust temporal behavior in a time series, thereby facilitating sequence classification, such as machine translation. As it can be seen in the Fig. 4. In details, it takes the x_0 from the sequence of input and then it outputs h_0 and x_1 is the input for the next step. Similarly, h_1 from the next is the input with x_2 for the next step and so on. This way, it keeps remembering the context while training.

$$h_t = f(W_{hx}x_t + W_{hh}h_{t-1} + b)$$
$$y_t = g(W_{yh}h_t + c)$$

Some errors backpropagate through time in general RNN. LSTM units was proposed to bypass these errors. While keeping a more consistent error, they let RNNs keep on learning over several time steps. LSTMs consist of information outside the basic flow of the rnn in a valved block. An LSTM cell displays in Fig. 3 and its functions are as below:

$$f_t = \sigma_g(W_f x_t + U_f h_{t-1} + b_f)$$
$$i_t = \sigma_g(W_i x_t + U_i h_{t-1} + b_f)$$
$$o_t = \sigma_g(W_o x_t + U_o h_{t-1} + b_o)$$
$$c_t = f_t o c_{t-1} + i_t o \sigma_c(W_c x_t + U_c h_{t-1} + b_c)$$
$$h_t = o_t o \sigma_h(c_t)$$

Neural network's notes get triggered by the notes they get. Likewise, LSTM's gates pass on or block the data based on its weight. After that, these signals are grated with their own sets of weights. Subsequently, RNN's learning process modify these weights that control hidden states and input. Ultimately, these cells learn when to let information to get in, get out or be deleted through the consequent steps of taking guesses, back propagating error, and modulating weights via gradient descent.

3.4   RCNN Model

We adopt a model based on recurrent convolutional neural networks(RCNN) in multi-label classification for Chinese recipes, which was displayed in Fig. 5. We take the advantage of the CNN's local feature extraction capability, and the temporal and long dependencies of labels and recipe ingredients are processed by LSTM(long short-term memory) networks, a popular variation of recurrent neural networks.

In RCNN model, we actually stack CNN model with the LSTM networks, we let the output of the convolutional neural networks be a single scalar, combined with the label for further processing. And CNN (Convolutional neural networks) are used to extracted local features of an image or any other vectorized datasets through its convolutional layers.

Meanwhile, convolution is an element wise matrix multiplication, which plays an key role in NLP tasks that different filters applied to the contents may lead to a variety of effects. As long as training continues, the filters are updated to reduce errors. The convolution layer includes activation functions, which are mainly nonlinear correction linear units.

We choose k-MaxPooling layers in between convolutional layers and BiLSTM encoder. After several combinations of convolutional and pooling layers, we have a final flattening layer to convert our 3d tensor into 1d array in order to feed into fully connected layers. We simply connect a matrix or tensor row by row into a flatten 1d array.

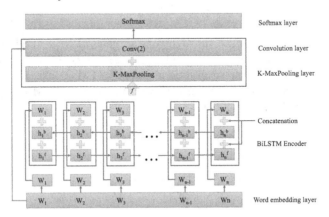

Fig. 5. RCNN model

The LSTM architecture is composed of a collection of recurrently connected chained networks, which be called as memory blocks. LSTM network contains memory cells and so-called 'gates', providing operations on the cells. These gates, input, output and the forget gates, are the most distinguishable part of the LSTM networks from the traditional RNNs. Typically, these gates are implemented as activation function like a sigmoid function in order to decide the amount of the information to be passed through.

An overview of the RCNNs model as shown in Fig. 5, which combining word embedding, CNN, and LSTM.

4 Experiments

4.1 Datasets and Setup

The recipe dataset we use here is collected from meishij, which is one of the biggest recipe websites in China. There are more than 15,747 recipes from various Chinese cuisines in the dataset which contain labels annotated by users. The vocabulary of words is 5,697 in the dataset, the vocabulary of labels is 447, and the average number of words and labels in each recipe is 7.88 and 7.51 respectively. The dataset has been splitted into training set (12,747 recipes) and test set (3,000 recipes). In our experiment, we randomly select 10% of the training set as the Validation set.

Data preprocessing work includes word embedding representation and relation labels numeralization. W is the pretrained word embedding matrix. It is updated as a model parameter in the optimization process. The Word2Vec tool training on the Skip-gram model is used as a pre-trained word embedding matrix, which contains 123,496 ingredient words. The dimension of word vector is 100.

$$Precision = \sum_{pos\ in\{1,2,3,4,5\}} \frac{Precision@pos}{log(pos + 2)}$$

$$Recall = \frac{Right\_label\_num}{All\_marked\_label\_num}$$

The labels in our recipe dataset are not duplicated by default. The top-5 labels of each recipe predicted by our model are sorted from large to small according to the their probability, then we discard the rest of labels which start from 6. In order to evaluate the performance of our models, we design precision, recall and score. For top-5 labels, hitting any positions of the real labels is considered true. We give the accuracy of each position a weight in terms of their position, then we obtain the precision. The recall is the quotient between $Right\_label\_num$ and $All\_marked\_label\_num$. Finally, the score is the harmonic mean between precision and recall.

$$Score = \frac{Precision * Recall}{Precision + Recall}$$

4.2   Baseline

In this section, we compare proposed deep neural network models with some baseline models. We consider the following methods.

- Naive Bayes: This algorithm belongs to the top 10 algorithms in data mining [18], which be used for the classification tasks. We used Naive Bayes classifier chains for the tag classification task, which transforms a multi-label problem into multiple binary problem. Given the ingredients of the recipes, we can estimate the posterior probability of each tag.
- MLKNN: we used the MLKNN [20] method for the tag classification, which is the multi-label version of KNN.
- fastText: We used fastText model as one of our baseline, which similar with the CBOW in word2vec. In this model, some of the n-gram features have been added to capture local sequence information.

Naive Bayes and MLKNN are traditional machine learning methods. And fastText is a quick text classification model, which is attracted a lot of attention in recent year.

4.3   Results and Discussion

Parameters Sensitive Analysis. In terms of the models we proposed, there are several hyper-parameters impact the performance. To evaluate their influence on CNNs, we evaluate two important ones, window sizes w and the number of filters n. For RNNs, we evaluate three ones, Bi-directional b, pooling ways p and hidden sizes h. Both of these models are evaluated whether to performance data augmentation.

Table 1. The effect of window size on CNN

Window size	Score	Precision	Recall
1	0.3971	3.0963	0.4555
2	0.4037	3.1534	0.4629
3	0.3701	2.9051	0.4242
1, 3	0.4096	3.1974	0.4698
1, 2, 3	**0.4138**	**3.2279**	**0.4747**

Table 1 shows the performance of various window sizes on our CNNs models. We use the following windows sizes settings: 1, 2, 3, (1, 3), (1, 2, 3). From the results of table, we got the best result when the window sizes were (1, 2, 3). And when window sizes equal to 1, 2, or 3, we can clearly see that the best window size is 2 on our multi-label classification task. The reason is that when the window size equals to 1, the convolutional operation will extract the unigram

Table 2. The effect of number of filter on CNN

Number of filter	Score	Precision	Recall
64	0.4049	3.1579	0.4645
100	0.3936	3.0784	0.4513
156	0.4058	3.1647	0.4655
200	0.4116	3.2119	0.4721
250	0.4138	3.2279	0.4747
300	0.4126	3.2193	0.4733

information while ignore the context information. The results of window sizes equal to 2 and 3 show that the bigram information is more important than the trigram for our task. From the results we also can observe that the multiple window sizes can achieve better performance than not multiple ones. This can help demonstrated that the advantages of the models with multiple window sizes over the single window size models. Comparing the results of window sizes equal to (1, 3) and (1, 2, 3), we can observe that the performance between the multiple window sizes were similar, so we can choose the multiple window size conveniently. Table 2 shows the performance of CNNs model with various number of filters. We set the number of filters with 64, 100, 156, 200, 250 and 300. From the result, we can observe that the best filters number is 250 on our multi-label classification task. Beside, when the number of filters from 250 to 300, there score are similar but have a little decrement. So we set the number of filter with 250 as proper parameter.

Table 3. The effect of pooling ways on RNN

Bi-directional	Pooling	Score	Precision	Recall
FALSE	2-MaxPooling	0.3402	2.6555	0.3902
TRUE	**2-MaxPooling**	**0.4017**	**3.1416**	**0.4605**
FALSE	3-MaxPooling	0.3603	2.8184	0.4132
TRUE	**3-MaxPooling**	**0.3898**	**3.0498**	**0.4469**
FALSE	4-MaxPooling	0.3535	2.7648	0.4053
TRUE	**4-MaxPooling**	**0.3943**	**3.0847**	**0.4521**

Table 3 represents the performance of different pooling ways for directional or bidirectional RNN models. The settings include 2-MaxPooling, 3-MaxPooling or 4-MaxPooling ways. After the experiments, we obtained the best performance when bi-directional RNN with 2-MaxPooling ways. From the result, we also can observe that bidirectional RNN obtained better performance than the directional one, that proved that bidirectional architecture can make full use of the

Table 4. The effect of hidden size on RNN

Hidden size	Score	Precision	Recall
64	0.3813	2.9813	0.4372
128	0.4011	3.1328	0.4601
200	**0.4017**	**3.1416**	**0.4605**
256	0.3742	2.9285	0.4289
320	0.3932	3.0734	0.4509

context of the textual contents. Meanwhile, Table 4 shows the performance of various hidden sizes for our RNN model. The hidden sizes with the following settings: 64, 128, 200, 256 and 320. From our experiments, we obtained the best performance when the hidden size with 200 in our RNN model. Besides, we clearly observed that the scores of our RNN model incrementing placidly when the hidden sizes from 64 to 200, and decrementing when from 200 to 320. This is because too many parameters will make the model overfit more easier, and weaken robustness. Meanwhile, setting too large hidden sizes will increase the training times, so the suitable parameter for our multi-label classification task is 200.

Table 5. The comparisons of RCNN with CNN and RNN

Method	Score	Precision	Recall
CNN	0.4138	3.2279	0.4747
RNN	0.4017	3.1416	0.4605
RCNN	**0.4227**	**3.2948**	**0.4849**

Based on parameters sensitive analysis for CNN and RNN above, we adjust our RCNN model with the following settings: the window sizes with (1, 2, 3), 250 filters, Bidirectional LSTM with 2-MaxPooling and the hidden size with 200. Table 5 shows the result of comparisons of RCNN with CNN and RNN. We can clearly observed that RCNN obtained higher score than another two models, which shown that RCNN can combine the advantages of CNN and RNN model.

The Effects of Data Augmentation. In order to estimate the effects of data augmentation in multi-label classification for Chinese recipes, we train the proposed models with data augmentation on our recipe training sets and test their performance on our recipe testing sets. In details, the data augmentation strategy we adapted including two operations process: random swap and random deletion.

Table 6 shows the performance of CNN, RNN and RCNN models with data augmentation or not. Obviously, we can observed that data augmentation can

Table 6. The effects of data augmentation

Models	No-data augmentation			With-data augmentation		
	Score	Precision	Recall	Score	Precision	Recall
CNN	0.4138	3.2279	0.4747	0.4150	3.2328	0.4761
RNN	0.4017	3.1416	0.4605	0.4088	3.1946	0.4689
RCNN	0.4227	3.2948	0.4849	0.4248	3.3111	0.4873

help improve the score of most proposed models. And RNN model had a higher promotion than others. The use of data augmentation can easily raise the score by expand the dataset.

Comparisons with Other Methods. Table 7 shows the comparisons of the results of different methods on the test sets, including the baseline methods and various state-of-art deep neural network models. "data augment" represent that specific model using data augmentation. From Table 7 we can clearly observed that deep neural networks are better than the baseline models, especially the RCNN model. "Naive Bayes" and "MLKNN" are based on the traditional feature extract process, others are based on word embedding. Meanwhile, for each test recipe data we first obtained Top-5 labels according to their predicted score by various deep neural networks and baselines models, then we calculated the accuracy of each position, where the details are shown in Fig. 6. From this figure we can clearly observed that segments of RCNN, CNN, RNN and fastText are very closed, Naive Bayes and MLKNN are in the bottom. For each model, the first position obtain the highest accuracy, and then the trends go down after it.

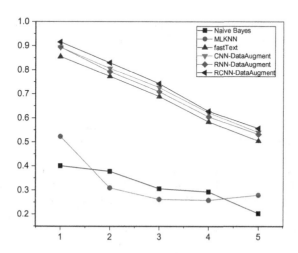

Fig. 6. Accuracy with the variation of predicted tag number

Table 7. The comparisons on different models

Method	Score	Precision	Recall
Naive Bayes	0.1829	1.4372	0.2097
MLKNN	0.1898	1.5405	0.2165
fastText	0.3942	3.0781	0.4521
CNN	0.4138	3.2279	0.4747
RNN	0.4017	3.1416	0.4605
RCNN	0.4227	3.2948	0.4849
CNN-data augment	0.4150	3.2328	0.4761
RNN-data augment	0.4088	3.1946	0.4689
RCNN-data augment	0.4248	3.3111	0.4873

5 Conclusion

In this paper, we empirically conduct our work by the deep neural networks in multi-label classification for Chinese recipes. For further comparisons, we also adopted Naive Bayes, MLKNN and fastText as our baseline models. Our experiments have shown that the performances of deep neural networks are better than the baseline models. Besides, RCNN model perform better than CNN and RNN, which made full use of the advantage of its architecture. Besides, data augmentation helped improve the performance of proposed models by expand the dataset. Meanwhile, for each test recipe data we obtained Top-5 labels according to their predicted scores by the proposed models, then calculated the accuracy of each position respectively.

Acknowledgment. This work was supported by the project of Natural Science Foundation of China (No. 61402329, No. 61972456) and the Natural Science Foundation of Tianjin (No. 19JCYBJC15400).

References

1. Cui, X., Goel, V., Kingsbury, B.: Data augmentation for deep neural network acoustic modeling. IEEE/ACM Trans. Audio Speech Lang. Process. **23**(9), 1469–1477 (2015)
2. Dembczynski, K., Cheng, W., Hüllermeier, E.: Bayes optimal multilabel classification via probabilistic classifier chains. In: ICML, pp. 279–286 (2010)
3. Dembszynski, K., Waegeman, W., Cheng, W., Hüllermeier, E.: On label dependence in multilabel classification. In: Proceedings of the LastCFP: ICML Workshop on Learning from Multi-Label Data, p. 8 (2010)
4. Ji, S., Tang, L., Yu, S., Ye, J.: A shared-subspace learning framework for multi-label classification. TKDD **4**(2), 8:1–8:29 (2010)
5. Joulin, A., Grave, E., Bojanowski, P., Mikolov, T.: Bag of tricks for efficient text classification. CoRR abs/1607.01759 (2016)

6. Kandola, E.J., Hofmann, T., Poggio, T., Shawe-Taylor, J.: A neural probabilistic language model. Stud. Fuzziness Soft Comput. **194**, 137–186 (2006)
7. Kim, Y.: Convolutional neural networks for sentence classification. CoRR abs/1408.5882 (2014)
8. Ko, T., Peddinti, V., Povey, D., Khudanpur, S.: Audio augmentation for speech recognition. In: INTERSPEECH, pp. 3586–3589 (2015)
9. Kobayashi, S.: Contextual augmentation: data augmentation by words with paradigmatic relations. CoRR abs/1805.06201 (2018)
10. Krizhevsky, A., Sutskever, I., Hinton, G.E.: ImageNet classification with deep convolutional neural networks. Commun. ACM **60**(6), 84–90 (2017)
11. Lai, S., Xu, L., Liu, K., Zhao, J.: Recurrent convolutional neural networks for text classification. In: Proceedings of the Twenty-Ninth AAAI Conference on Artificial Intelligence, pp. 2267–2273 (2015)
12. Liu, P., Qiu, X., Huang, X.: Recurrent neural network for text classification with multi-task learning. CoRR abs/1605.05101 (2016)
13. Read, J., Pfahringer, B., Holmes, G., Frank, E.: Classifier chains for multi-label classification. Mach. Learn. **85**(3), 333–359 (2011). https://doi.org/10.1007/s10994-011-5256-5
14. Rong, X.: word2vec parameter learning explained. CoRR abs/1411.2738 (2014)
15. Szegedy, C., et al.: Going deeper with convolutions. In: IEEE Conference on Computer Vision and Pattern Recognition, CVPR, pp. 1–9 (2015)
16. Tsoumakas, G., Katakis, I., Vlahavas, I.P.: Random k-labelsets for multilabel classification. IEEE Trans. Knowl. Data Eng. **23**(7), 1079–1089 (2011)
17. Wei, J.W., Zou, K.: EDA: easy data augmentation techniques for boosting performance on text classification tasks. CoRR abs/1901.11196 (2019)
18. Wu, X., et al.: Top 10 algorithms in data mining. Knowl. Inf. Syst. **14**(1), 1–37 (2008). https://doi.org/10.1007/s10115-007-0114-2
19. Xia, X., Yang, X., Li, S., Wu, C., Zhou, L.: RW.KNN: a proposed random walk KNN algorithm for multi-label classification. In: Proceedings of the 4th Workshop on Workshop for Ph.D. Students in Information & Knowledge Management, pp. 87–90 (2011)
20. Zhang, M., Zhou, Z.: ML-KNN: a lazy learning approach to multi-label learning. Pattern Recogn. **40**(7), 2038–2048 (2007)

Improving Word Alignment with Contextualized Embedding and Bilingual Dictionary

Minhan Xu[ID] and Yu Hong[(✉)]

Soochow University, Suzhou 215006, Jiangsu, China

Abstract. Word alignment is a natural language processing task that identifies the relationship of the among words of multiword units in a bitext. Large pre-trained models can generate significantly improved contextual word embedding. However, Statistical methods are still preferred choices. In this paper, we utilize bilingual dictionaries and contextualized word embeddings generated by pre-trained models in word alignment. We use statistical methods to generate rough alignment first, then use bilingual dictionaries to modify the alignment to make it more accurate. We use this alignment as training data and leverage this in training to optimize alignment. We demonstrate that our approach produces better or comparable performance compared to statistical approaches.

Keywords: Natural language processing · Word alignment · Word embedding

1 Introduction

Machine translation is proposed to model the semantic equivalence between a pair of source and target sentences [11]. Similarly, Word alignment tries to model the semantic equivalence between a pair of source and target words [21]. Word alignment is closely related to machine translation and such a relation can be traced back to the birth of statistical machine translation (SMT) [2]. In SMT, Word alignment is the basis of SMT models, because parameters of SMT models are typically estimated by observing word-aligned bitexts, conversely, automatic word alignment is typically done by choosing the alignment which best fits an SMT model and its accuracy is generally helpful to improve translation quality [13,15].

With the rise of neural machine translation (NMT) [1], researchers have been attempting to extract word alignments from attention matrices [10,12]. Several methods create alignments from attention matrices [14,24,28] or pursue a multitask approach for alignment and translation [3,9].

Nowadays, word alignment using statistical methods is still a go-to solution. Statistical methods such as fast-align [7], GIZA++ [21], eflomal [23] work well, while over the past several years, neural networks have already taken the

H. Mei et al. (Eds.): BigData 2020, CCIS 1320, pp. 180–194, 2021.
https://doi.org/10.1007/978-981-16-0705-9_13

lead in many natural language processing tasks. Many pre-trained models are proposed to generate contextualized word embedding such as BERT, OpenAI GPT, Transformer-XL, XLM, XLNet, ALBERT, RoBERTa. These pre-trained models use massive corpora to train, thus generate excellent contextual word representations.

In this paper, we present a method to utilize these contextual word representations and bilingual dictionaries to help improving alignment generated from statistical word alignment approaches. First, we extract the vocabulary in the pre-trained BERT model in both languages, then we find the corresponding word in both vocabularies using a bilingual dictionary, next, we can achieve the token id mapping by applying respective tokenizer. After this, we apply the statistical word alignment methods to generate the rough alignment, at last, we use synonym token id mapping to modify the alignment and treat this as training data for our model. We find that our approach shows better or comparable performance compared to statistical approaches.

2   Preliminaries

2.1   Word Alignment Task

The task of word alignment consists of finding correspondences between words and phrases in parallel texts. Assuming a sentence aligned bilingual corpus in languages L_s and L_t, the task of a word alignment system is to indicate which word token in the corpus of language L_s corresponds to which word token in the corpus of language L_t.

Given a sentence in source language $S = s_1, s_2, ..., s_m$ and a sentence in target language $T = t_1, t_2, ..., t_n$, an alignment is defined as a subset of the Cartesian product of the word positions [21].

$$A \subseteq \{(i,j) : i = 1, \ldots, m; j = 1, \ldots, n\} \tag{1}$$

In the alignment, there may be some words in language L_s map to multiple words in language L_t.

2.2   BERT Model

BERT (Bidirectional Encoder Representations from Transformers) is a model presented by Google, it makes use of Transformer, an attention mechanism that learns contextual relations between words (or sub-words) in a text. A vanilla transformer model contains two separate parts—an encoder that reads the text input and a decoder that produces a prediction for the task. Since BERT's goal is to generate a language model, only the encoder mechanism is necessary. As opposed to directional models, which read the text input sequentially (left-to-right or right-to-left), the Transformer encoder reads the entire sequence of words at once. Therefore it is considered bidirectional. This characteristic allows the model to learn the context of a word based on all of its surroundings (left and right of the word).

BERT is designed to pre-train deep bidirectional representations from unlabeled text by jointly conditioning on both left and right context. As a result, the pre-trained BERT model can be fine-tuned with just one additional output layer to create state-of-the-art models for a wide range of NLP tasks. BERT embeds the input sentence with the sum of the token embeddings, the segmentation embeddings and the position embeddings. They use two strategies for training BERT, the first is mask language model (MLM), by masking out some of the words in the input and then condition each word bidirectionally to predict the masked words. Before feeding word sequences into BERT, 15% of the words in each sequence are replaced with a [MASK] token. The model then attempts to predict the original value of the masked words, based on the context provided by other non-masked words in the sequence.

The second technique is the next sentence prediction (NSP), where BERT learns to model relationships between sentences. In the training process, the model receives pairs of sentences as input and learns to predict if the second sentence in the pair is the subsequent sentence in the original document. For example:

Sentence A : My phone fell to the ground.
Sentence B : It couldn't start.
Label : IsNextSentence

Sentence A : My phone fell to the ground.
Sentence B : The apple is so delicious.
Label : NotNextSentence

Fig. 1. Example of next sentence prediction

Intuitively, it is reasonable to believe that a deep bidirectional model is strictly more powerful than either a left-to-right model or the shallow concatenation of a left-to-right and a right-to-left model. That's where BERT greatly improves upon both GPT and ELMo.

The BERT architecture builds on top of Transformer. There are two variants available: BERT Base has 12 layers (transformer blocks), 12 attention heads and 110 million parameters; BERT Large has 24 layers, 16 attention heads and 340 million parameters. On SQuAD v1.1, BERT achieves 93.2% F1 score, surpassing the previous state-of-the-art score of 91.6% and human-level score of 91.2%: BERT also improves the state-of-the-art by 7.6% absolute on the very challenging GLUE benchmark, a set of 9 diverse Natural Language Understanding (NLU) tasks.

2.3 IBM Alignment Models

IBM alignment models are a sequence of increasingly complex models used in statistical machine translation to train a translation model and an alignment model, starting with lexical translation probabilities and moving to reordering and word duplication. They underpinned the majority of statistical machine translation systems for almost twenty years starting in the early 1990s, until neural machine translation began to dominate.

IBM Model 1 is weak in terms of conducting reordering or adding and dropping words. In most cases, words that follow each other in one language would have a different order after translation, but IBM Model 1 treats all kinds of reordering as equally possible. Another problem while aligning is the fertility (the notion that input words would produce a specific number of output words after translation). In most cases one input word will be translated into one single word, but some words will produce multiple words or even get dropped (produce no words at all). The fertility of word models addresses this aspect of translation.

The IBM Model 2 has an additional model for alignment that is not present in Model 1. For example, using only IBM Model 1 the translation probabilities for these translations would be the same. The IBM Model 2 addressed this issue by modeling the translation of a foreign input word in position i to a native language word in position j using an alignment probability distribution defined as:

$$a\left(i \vee j, l_e, l_f\right) \tag{2}$$

In the above equation, the length of the input sentence f is denoted as l_f, and the length of the translated sentence e as l_e. The translation done by IBM Model 2 can be presented as a process divided into two steps: lexical translation and alignment. Assuming $t(e \mid f)$ is the translation probability and $a(i \vee j, l_e, l_f)$ is the alignment probability, IBM Model 2 can be defined as:

$$p(e, a \mid f) = \prod_{j=1}^{l_e} t(e_j \vee f_{a|j}) a(a(j) \vee j, l_e, l_f) \tag{3}$$

In this equation, the alignment function a maps each output word j to a foreign input position $a(j)$.

The fertility problem is addressed in IBM Model 3. The fertility is modeled using probability distribution defined as:$n(\phi \vee f)$. For each foreign word j, such distribution indicates to how many output words it usually translates. This model deals with dropping input words because it allows $\phi = 0$. But there is still an issue when adding words. For example, the English word do is often inserted when negating. This issue generates a special $NULL$ token that can also have its fertility modeled using a conditional distribution defined as: $n(\varnothing \vee NULL)$. The number of inserted words depends on sentence length. This is why the $NULL$ token insertion is modeled as an additional step: the fertility step. It increases the IBM Model 3 translation process to four steps: fertility step, $NULL$ insertion step, lexical translation step and distortion step. The last step is called distortion instead of alignment because it is possible to produce the same translation with the same alignment in different ways.

In IBM Model 4, each word is dependent on the previously aligned word and on the word classes of the surrounding words. Some words tend to get reordered during translation more than others (e.g. adjective–noun inversion when translating Polish to English). Adjectives often get moved before the noun that precedes them. The word classes introduced in Model 4 solve this problem by conditioning the probability distributions of these classes. The result of such distribution is a lexicalized model. Such a distribution can be defined as follows: $d_1(j - \odot_{[i-1]} \vee A(f_{[i-1]}), B(e_j))$ for the initial word in the cept and $d_1(j - \pi_{i,k-1} \vee B(e_j))$ for additional words. where $A(f)$ and $B(e)$ functions map words to their word classes, and e_j and $f_{[i-1]}$ are distortion probability distributions of the words. The cept is formed by aligning each input word f_i to at least one output word. Both Model 3 and Model 4 ignore if an input position was chosen and if the probability mass was reserved for the input positions outside the sentence boundaries. It is the reason for the probabilities of all correct alignments not sum up to unity in these two models.

IBM Model 5 reformulates IBM Model 4 by enhancing the alignment model with more training parameters in order to overcome the model deficiency. During the translation in Model 3 and Model 4 there are no heuristics that would prohibit the placement of an output word in a position already taken. In Model 5 it is important to place words only in free positions. It is done by tracking the number of free positions and allowing placement only in such positions. The distortion model is similar to IBM Model 4, but it is based on free positions. If v_j denotes the number of free positions in the output, the IBM Model 5 distortion probabilities would be defined as: $d_1(v_j \vee B(e_j), v_{\odot i-1}, v_{max})$ for the initial word in the cept and $d_1(v_j - v_{\pi_{i,k-1}} \vee B(e_j), v_{max'})$ for additional words. The alignment models that use first-order dependencies like the HMM or IBM Models 4 and 5 produce better results than the other alignment methods. The main idea of HMM is to predict the distance between subsequent source language positions. On the other hand, IBM Model 4 tries to predict the distance between subsequent target language positions.

2.4   Bayesian IBM Models

The IBM models make no a priori assumptions about the categorical distributions that define the model, and most authors have used maximum-likelihood estimation through the Expectation-Maximization algorithm [5] or some approximation to it. However, when translating natural languages the lexical distributions should be very sparse, reflecting the fact that a given source word tends to have a rather small number of target words as allowable translations, while the vast majority of target words are unimaginable as translations. These constraints have recently been modeled with sparse and symmetric Dirichlet priors [17,18,26], which, beyond capturing the range of lexical distributions we consider likely, also turn out to be mathematically very convenient as the Dirichlet distribution is a conjugate prior to the categorical distribution. Another direction of research has explored hierarchical distributions such as the Pitman-Yor process [25] instead of the Dirichlet distribution for the translation distribution priors [8,22]. Such distributions offer even greater flexibility in specifying prior constraints on the categorical distributions, but at the cost of less efficient inference.

2.5   EFLOMAL and EFMARAL

eflomal is short for Efficient Low-Memory Aligner. It is a word alignment tool based on efmaral [23], with the following main differences: more compact data structures are used, so memory requirements are much lower, the estimation of alignment variable marginals is done one sentence at a time, which also saves a lot of memory at no detectable cost in accuracy.

efmaral is a new system for efficient and accurate word alignment using a Bayesian model with Markov Chain Monte Carlo (MCMC) inference. It managed to surpass the fast_align system which is commonly used for performance-critical word alignment, both in computational efficiency and alignment accuracy. In efmaral model, MCMC, in particular Gibbs sampling is used for inference in IBM alignment models. Here is the brief summary: Given a probability function $P_M(x)$ of some model M on parameter vector x, MCMC provides us with the means to draw samples from P_M. This is done by constructing a Markov chain with values of x as states, such that its stationary distribution is identical to P_M. In practice, this means deriving expressions for the transition probabilities $P(x' \mid x)$ of going from state x to state x'. Since the number of states is enormous or infinite in typical applications, it is essential that there is some way of sampling efficiently from $P(x' \mid x)$. With Gibbs sampling, this is done by sampling one variable from the parameter vector x at a time, conditioned on all other variables: $P(x_i \mid x_1, x_2, \ldots, x_{i-1}, x_{i+1}, \ldots, x_m)$ to indicate conditioning on all elements of x except at index i. All positions i are then sampled in some arbitrary but fixed order. By choosing suitable distributions for the model, the goal in designing a Gibbs sampler is to make sure that this distribution is easy to sample from.

3   Proposed Method

3.1   Word Alignment as Multi-class Classification

In this paper, we treat the word alignment task as a multi-class classification task and fine-tune the pre-trained BERT model to turn it into an alignment model. First, we extract the vocabulary V_s^m and V_t^n in the pre-trained BERT model in both source and target languages. Then we find the corresponding word in both vocabularies using a bilingual dictionary D. In this process, we may be faced with a many-to-many word mapping situation. After this, we will get a word mapping dictionary D',

$$D' = \{(w_s^i, w_t^j), \ldots : i = 1, \ldots, m; j = 1, \ldots, n\} \tag{4}$$

Next, we achieve the token id mapping M by applying the respective tokenizer to the word mapping. In training, we input a source language sentence S and a target language sentence T to respective pre-trained BERT model. We combine the subword embedding to word embedding by adding them up. Then we calculate the similarity between the source language and target language using matrix multiplication. We also tried applying cosine similarity but it didn't work well. Besides calculating word similarity, we also add a neural network called *output_num_layer* after BERT model to solve the many-to-many alignment problem. There are two different *output_num_layer* for the respective language. In this layer, there are three sets of linear layer and ReLU activation layer and at last a linear layer to output a number. This number represents how many words should be in its alignment. If the number is zero, it means it doesn't need to be aligned.

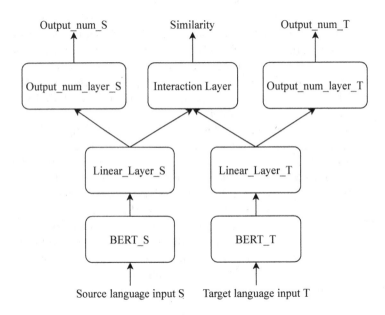

Fig. 2. Structure of our model

3.2   Training

We apply the statistical word alignment methods to generate the rough alignment A, then, we use synonym token id mapping M to modify the alignment and treat this as training data for our model. We modify the alignment by this rule: for any word w_s in source language, we get the corresponding translation w_t according to token id mapping M, then we count the frequency of occurrence of w_t in the target language sentence T, if w_t only appears once, we replace the word alignment with w_s and w_t. We use cross-entropy loss function for multiclass classification and mean squared error for *output_num_layer*.

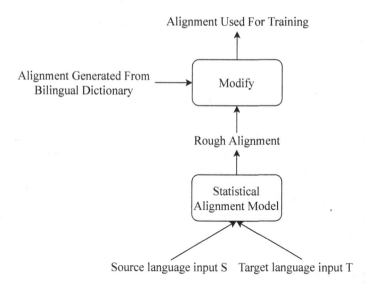

Fig. 3. Generate Training Data

Our model has three losses, namely the multi-class classification loss and *output_num_layer* losses for separate languages.

$$loss_a = -\sum_i \log\left(\frac{\exp(x_i[class])}{\sum_j \exp(x_i[j])}\right) \tag{5}$$

$$loss_b = \frac{1}{m}\sum_{t=1}^{m}\left(x_{num}^t - x_{num}^{t'}\right)^2 \tag{6}$$

$$loss_c = \frac{1}{n}\sum_{t=1}^{n}\left(y_{num}^t - y_{num}^{t'}\right)^2 \tag{7}$$

In our experiment, we used BERT-Base-Uncased [6] (12-layer, 768-hidden, 12-heads) for English, CamemBERT-Base [16] (12-layer, 768-hidden, 12-heads) for French, BERT-Base-Romanian-Uncased-v1 (12-layer, 768-hidden, 12-heads) for Romanian and Hindi-BERT (12-layer, 256-hidden, 12-heads) trained by Nick Doiron for Hindi. The following are the parameters: learning rate = 1e-5, max seq length = 512, max query length = 160, truncation = True.

3.3 Inference

According to the number x_{num} and y_{num} generated from *output_num_layer*, on source language side, we choose the top x_{num} most similar words in similarity matrix as the alignment and the top y_{num} most similar words in similarity matrix as the alignment on target language side. When x_{num} or y_{num} is zero, we omit the alignment. We tested the inference from both directions, that is from source language to target language and from target language to source language. Also, we tried to take the intersection and union of these two alignment sets.

3.4 Baselines

We use two popular statistical alignment models which are eflomal and GIZA++.

Eflomal (based on efmaral) is a Bayesian model with Markov Chain Monte Carlo inference, is claimed to outperform fast-align on speed and quality. Eflomal creates forward and backward alignments and then symmetrize them. We generate the symmetrization result using grow-diagfinal-and (GDFA) method. GIZA++ is the most widely used word alignment toolkit. It combines IBM Alignment Models. We use its standard settings: 5 iterations each for the HMM model, IBM Model 1, 3 and 4 with $p_0 = 0.98$. Experiments are all done on word level with statistical model.

4 Experiments

4.1 Data

We use bilingual dictionaries from MUSE [4] and three language pairs from the WPT2005 shared task: English-Hindi, English-French [21], and English-Romanian [19]. Note that English-Hindi and English-Romanian test data only contain sure edges and no possible edges, while English-French test data contain both sure edges and possible edges. We will treat all the alignment result generated from our model as sure edges because our model can't produce possible edges. Data statistics are shown in Table 1.

Table 1. Number of sentences for three datasets:English-French (ENG-FRA), English-Romanian (ENG-RON), English-Hindi (ENG-HIN).

Language	Gold standard size	Parallel data size
ENG-FRA	447	1130104
ENG-RON	248	48441
ENG-HIN	90	3441

4.2 Evaluation Measures

Given a set of predicted alignment edges A and a set of sure (possible) gold standard edges S (P), we use the following evaluation measures:

$$\text{prec.} = \frac{|A \cap P|}{|A|} \qquad (8)$$

$$\text{rec.} = \frac{|A \cap S|}{|S|} \tag{9}$$

$$F_1 = \frac{2prec.rec.}{prec. + rec.} \tag{10}$$

$$\text{AER} = 1 - \frac{|A \cap S| + |A \cap P|}{|A| + |S|} \tag{11}$$

We use the perl script provided by WPT2005 and the nonullalign version of the alignment result for test data throughout the paper.

5   Results and Discussion

Table 2 and 3 compare the performance of our methods against statistical baselines using three language pairs. The improvement of our method suggests that pre-trained models combined with bilingual dictionaries help produce better word alignment.

In the table, our model plus GIZA++ method means we use GIZA++ generated training data to train our model and our model plus eflomal means we user eflomal generated training data. L1-L2 means we infer the alignment from left language to right language and vice versa. Intersection means we take the intersection of the L1-L2 and L2-L1 alignment. Union means we take the union of the L1-L2 and L2-L1 alignment.

Table 2. Results on the English-French alignment task with bilingual dictionary

Method	ENG-FRA						
	Sure align			Probable Align			AER
	Prec.	Rec.	F1	Prec.	Rec.	F1	
GIZA++	50.0	86.4	63.3	81.2	32.5	46.4	16.9
eflomal	56.1	90.9	69.4	88.1	33.0	48.1	10.8
our model+GIZA++ L1-L2	53.9	93.3	68.3	85.8	34.4	49.1	11.5
our model+GIZA++ L2-L1	54.4	90.3	67.9	88.2	33.9	49.0	11.0
our model+GIZA++ intersection	64.8	87.9	74.6	94.9	29.8	45.4	8.1
our model+GIZA++ union	47.0	95.6	63.0	81.7	38.5	52.3	13.7
our model+eflomal L1-L2	54.7	94.1	69.2	88.0	35.1	50.2	9.8
our model+eflomal L2-L1	56.1	90.5	69.2	89.2	33.3	48.5	10.4
our model+eflomal intersection	65.5	88.6	75.3	95.7	30.0	45.7	**7.3**
our model+eflomal union	48.4	96.0	64.4	83.7	38.4	52.6	12.2

As is shown in the table, on English-French dataset, our model achieved lower AER than the statistical method except the union method with eflomal. The intersection operation gets low recall but high precision while the union operation gets high recall but low precision which results in lower AER. Intersection method works best on the English-French dataset which achieves 7.3% AER. Our model can get lower AER using alignment generated by eflomal for training than those generated by GIZA++. The table shows that when train our model with better training data it generate better alignment. On English-French alignment task, inference from English to French and inference from French to English get similar result. Eflomal generated training data gets higher performance when inferring from English to French than from French to English. GIZA++ generated training data gets higher performance when inferring from French to English than From English to French.

Table 3. Results on the English-Romanian and English-Hindi alignment task with bilingual dictionary

Method	ENG-RON				ENG-HIN			
	Sure align			AER	Sure align			AER
	Prec.	Rec.	F1		Prec.	Rec.	F1	
GIZA++	75.7	60.9	67.5	32.5	49.6	43.5	46.4	53.6
eflomal	84.1	66.3	74.1	**25.9**	61.5	43.7	51.1	48.9
our model+GIZA++ L1-L2	79.1	58.4	67.2	32.8	54.3	45.1	49.3	50.7
our model+GIZA++ L2-L1	78.6	61.9	69.2	30.8	54.1	44.8	49.0	51.0
our model+GIZA++ intersection	89.7	53.1	66.7	33.3	54.6	44.7	49.1	50.9
our model+GIZA++ union	72.0	67.3	69.5	30.5	53.9	45.1	49.1	50.9
our model+eflomal L1-L2	82.9	66.9	74.1	**25.9**	64.2	45.1	53.0	47.0
our model+eflomal L2-L1	82.8	66.3	73.7	26.4	64.3	45.4	53.2	**46.8**
our model+eflomal intersection	91.2	62.2	73.9	26.1	64.7	45.1	53.1	46.9
our model+eflomal union	76.7	71.1	73.8	26.2	63.8	45.4	53.1	46.9

On English-Romanian dataset, the test dataset only contain sure alignment edges. The lowest AER in our method which is inference from English to Romanian is equal to eflomal's result. As is shown in the table, our model plus GIZA++ achieve lower AER with inference from Romanian to English while our model plus eflomal achieve lower AER with inference from English to Romanian.

On English-Hindi dataset, the test dataset only contain sure alignment edges. Because English-Hindi dataset only contain 3441 parallel sentences, bilingual dictionary works best on this dataset. Our method achieves two% lower AER than the original statistical alignment method.

On all of three datasets, eflomal outperformed GIZA++, as a result, our model trained with eflomal generated alignment outperformed that with GIZA++ generated alignment. Our method outperformed the statistical methods except in ENG-RON dataset.

We also tried to align words without modifying the alignment generated by statistical models. Experiment results are shown in Table 4.

Table 4. Results on the English-French alignment task without bilingual dictionary

Method	ENG-FRA	ENG-RON	ENG-HIN
	AER		
GIZA++	16.9	32.5	53.6
eflomal	10.8	**25.9**	48.9
our model+GIZA++ L1->L2	12.6	33.0	54.5
our model+GIZA++ L2->L1	11.0	30.9	54.2
our model+GIZA++ intersection	7.9	33.1	54.2
our model+GIZA++ union	14.8	31.1	54.5
our model+eflomal L1->L2	10.1	26.2	47.7
our model+eflomal L2->L1	10.4	26.6	47.8
our model+eflomal intersection	**6.8**	26.4	**47.6**
our model+eflomal union	12.9	26.4	47.9

It seems that modifying the alignment generated by statistical alignment with bilingual dictionaries is helpful for making better alignment. Our method with alignment modification outperformed the model without modification except the intersection method on English-French dataset. The reason might be the abundance of parallel sentences in English-French dataset. Bilingual dictionary is not so helpful on English-French dataset is mostly because there are over a million pairs of parallel sentences used in training English-French word alignment in statistical methods. The large amount of parallel data helps GIZA++ and eflomal know synonyms in both languages very well. Our method works best on English-Hindi dataset because it has only 3441 parallel sentences for training alignment model. Through the observation of the experimental result, we found that contextual word embedding and bilingual dictionary is especially helpful for alignment task when there's a lack of training data for the specific language.

6 Related Work

Garg et al. [9] proposed an unsupervised method that jointly optimizes translation and alignment objectives. They extract discrete alignments from the attention probabilities learnt during regular neural machine translation model training and leverage them in a multi-task framework to optimize towards translation and alignment objectives. They achieved a significantly better alignment error rate (AER) than GIZA++ when they supervised their model using the alignments obtained from GIZA++. Their model requires about a million parallel sentences for training the Transformer model. Similar to our model, they also utilized the alignment generated from GIZA++.

Recently, Sabet et al. [27] proposed an unsupervised method based on contextualized word embedding for word alignment. The key idea is to leverage multilingual word embeddings, both static and contextualized for word alignment. They take the output of the mBERT (pretrained on the 104 languages with the largest Wikipedias) or XLM-RoBERTa, and then calculate the cosine similarity between contextualized embeddings (assuming aligned words are close in embedding space). They used three approaches to extract alignment from the similarity matrix and two post-processing methods to modify the alignment. This method outperform statistical word aligners in six languages. The results are shown in Table 5:

Table 5. Results of mBERT And XLM-R on the ENG-FRA, ENG-RON and ENG-HIN alignment task

	ENG-FRA		ENG-RON		ENG-HIN	
Method	F1	AER	F1	AER	F1	AER
mBERT - Argmax	0.94	0.06	0.64	0.36	0.53	0.47
XLM-R - Argmax	0.93	0.06	0.70	0.30	0.58	0.42

Nagata et al. [20] treat the alignment task as a question answering task by formalize a word alignment problem as a collection of independent predictions from a token in the source sentence to a span in the target sentence. They finetuned the multilingual BERT on a manually created gold word alignment dataset. They significantly outperformed previous supervised and unsupervised word alignment methods without using any bitexts for pretraining.

7   Conclusion

We presented a word alignment method using pre-trained BERT and bilingual dictionaries based on statistical models generated alignment and achieved better or comparable performance in word alignment. We found that contextual word embedding and bilingual dictionary is especially helpful for alignment task when there's a lack of training data for the specific language. The future work includes how to make the similarity function between source and target word embedding more precise, redesigning the *output_num_layer* to handle the many to many alignment problem in a better way and improving the recall score.

References

1. Bahdanau, D., Cho, K., Bengio, Y.: Neural machine translation by jointly learning to align and translate. arXiv preprint arXiv:1409.0473 (2014)
2. Brown, P.F., Della Pietra, S.A., Della Pietra, V.J., Mercer, R.L.: The mathematics of statistical machine translation: parameter estimation. Comput. Linguist. **19**(2), 263–311 (1993)
3. Chen, W., Matusov, E., Khadivi, S., Peter, J.T.: Guided alignment training for topic-aware neural machine translation. arXiv preprint arXiv:1607.01628 (2016)
4. Conneau, A., Lample, G., Ranzato, M., Denoyer, L., Jégou, H.: Word translation without parallel data. arXiv preprint arXiv:1710.04087 (2017)
5. Dempster, A.P., Laird, N.M., Rubin, D.B.: Maximum likelihood from incomplete data via the EM algorithm. J. R. Stat. Soc. Ser. B (Methodol.) **39**(1), 1–22 (1977)
6. Devlin, J., Chang, M.W., Lee, K., Toutanova, K.: Bert: Pre-training of deep bidirectional transformers for language understanding. arXiv preprint arXiv:1810.04805 (2018)
7. Dyer, C., Chahuneau, V., Smith, N.A.: A simple, fast, and effective reparameterization of IBM model 2. In: Proceedings of the 2013 Conference of the North American Chapter of the Association for Computational Linguistics: Human Language Technologies, pp. 644–648 (2013)
8. Gal, Y., Blunsom, P.: A systematic Bayesian treatment of the IBM alignment models. In: Proceedings of the 2013 Conference of the North American Chapter of the Association for Computational Linguistics: Human Language Technologies, pp. 969–977 (2013)
9. Garg, S., Peitz, S., Nallasamy, U., Paulik, M.: Jointly learning to align and translate with transformer models. arXiv preprint arXiv:1909.02074 (2019)
10. Ghader, H., Monz, C.: What does attention in neural machine translation pay attention to? arXiv preprint arXiv:1710.03348 (2017)
11. Koehn, P.: Statistical Machine Translation. Cambridge University Press, Cambridge (2009)
12. Koehn, P., Knowles, R.: Six challenges for neural machine translation. arXiv preprint arXiv:1706.03872 (2017)
13. Koehn, P., Och, F.J., Marcu, D.: Statistical phrase-based translation. UNiversity Of Southern California Marina Del Rey Information Sciences Inst, Technical Report (2003)
14. Li, X., Liu, L., Tu, Z., Shi, S., Meng, M.: Target foresight based attention for neural machine translation. In: Proceedings of the 2018 Conference of the North American Chapter of the Association for Computational Linguistics: Human Language Technologies (Long Papers), vol. 1, pp. 1380–1390 (2018)
15. Liu, Y., Liu, Q., Lin, S.: Log-linear models for word alignment. In: Proceedings of the 43rd Annual Meeting of the Association for Computational Linguistics (ACL 2005), pp. 459–466 (2005)
16. Martin, L., et al.: CamemBERT: a tasty French language model. In: Proceedings of the 58th Annual Meeting of the Association for Computational Linguistics (2020)
17. Mermer, C., Saraçlar, M.: Bayesian word alignment for statistical machine translation. In: Proceedings of the 49th Annual Meeting of the Association for Computational Linguistics: Human Language Technologies, pp. 182–187 (2011)
18. Mermer, C., Saraçlar, M., Sarikaya, R.: Improving statistical machine translation using bayesian word alignment and gibbs sampling. IEEE Trans. Audio Speech Lang. Proces. **21**(5), 1090–1101 (2013)

19. Mihalcea, R., Pedersen, T.: An evaluation exercise for word alignment. In: Proceedings of the HLT-NAACL 2003 Workshop on Building and Using Parallel Texts: Data Driven Machine Translation and Beyond, pp. 1–10 (2003)

20. Nagata, M., Katsuki, C., Nishino, M.: A supervised word alignment method based on cross-language span prediction using multilingual bert. arXiv preprint arXiv:2004.14516 (2020)

21. Och, F.J., Ney, H.: A systematic comparison of various statistical alignment models. Comput. Linguist. **29**(1), 19–51 (2003)

22. Östling, R.: Bayesian models for multilingual word alignment. Ph.D. thesis, Department of Linguistics, Stockholm University (2015)

23. Östling, R., Tiedemann, J.: Efficient word alignment with markov chain monte carlo. Prague Bull. Math. Linguist. **106**(1), 125–146 (2016)

24. Peter, J.T., Nix, A., Ney, H.: Generating alignments using target foresight in attention-based neural machine translation. Prague Bull. Math. Linguist. **108**(1), 27–36 (2017)

25. Pitman, J., Yor, M.: The two-parameter poisson-dirichlet distribution derived from a stable subordinator. Ann. Probab. 855–900 (1997)

26. Riley, D., Gildea, D.: Improving the IBM alignment models using variational bayes. In: Proceedings of the 50th Annual Meeting of the Association for Computational Linguistics (Short Papers), vol. 2, pp. 306–310 (2012)

27. Sabet, M.J., Dufter, P., Schütze, H.: SimAlign: high quality word alignments without parallel training data using static and contextualized embeddings. arXiv preprint arXiv:2004.08728 (2020)

28. Zenkel, T., Wuebker, J., DeNero, J.: Adding interpretable attention to neural translation models improves word alignment. arXiv preprint arXiv:1901.11359 (2019)

Hypernetwork Model Based on Logistic Regression

Lei Meng[1,2,3], Zhonglin Ye[1,2,3], Haixing Zhao[1,2,3($\boxtimes$)], Yanlin Yang[1,2,3], and Fuxiang Ma[1]

[1] College of Computer, Qinghai Normal University, Xining 810001, Qinghai, China
h.x.zhao@163.com
[2] Key Laboratory of Tibetan Information Processing, Ministry of Education, Xining 810008, Qinghai, China
[3] Tibetan Information Processing and Machine Translation Key Laboratory of Qinghai Province, Xining 810008, Qinghai, China

Abstract. The evolution of hypernetworks is mostly based on growth and preferential connection. During the construction process, the number of new nodes and hyperedges is increasing infinitely. However, in the network, considering the impact of real resources and environment, the nodes can not grow without the upper limit, and the number of connections can not grow without the upper limit. Under certain conditions, there will be the optimal or maximum growth number. In addition, with the increase of the size of the hypernetwork and the number of nodes, there will be more old nodes waiting to be selected. Based on the problems in the process of constructing the hypernetwork, this paper improves the construction of the hypernetwork, and proposes a hypernetwork model construction method based on logical regression. Firstly, the maximum growth number of nodes is set to limit its growth, and the maximum capacity is set for each hyperedge; Secondly, the connection between nodes is selected by using logical regression instead of preferential connection; Finally, the old nodes are selected by using sub linear growth function instead of constant. Through the simulation experiment, it is found that the hyperdegree distribution of the improved hypernetwork model conforms to the power-law distribution; by changing the network scale, hyperedge capacity, the number of new nodes added and the number of old nodes selected in the process of the construction of the hypernetwork, the change law of the hypernetwork model's hyperdegree distribution is studied.

Keywords: Complex network · Hypernetwork model · Hyperdegree power law distribution · Logistic regression

1 Introduction

The rise and rapid development of complex networks provide a new direction for us to better study the network in the real world, and let us have a better understanding of the real world. Since Watts-Strogatz model [1] and Barabási-Albert model [2] were put forward, the research on complex network has been deepened and its research field is

© Springer Nature Singapore Pte Ltd. 2021
H. Mei et al. (Eds.): BigData 2020, CCIS 1320, pp. 195–208, 2021.
https://doi.org/10.1007/978-981-16-0705-9_14

also expanding. In recent years, as a tool to describe complex natural science and social system, complex network has been widely used in traffic network, biological network, scientific research cooperation network and social network. However, up to now, there is no uniform definition of complex network. Qian Xueshen gives a more strict definition of complex networks: networks with some or all of the properties of self-organization, self similarity, attractor, small world and scale-free are called complex networks. The study of complex network is actually the study of its topological structure. The topological structure of complex network is the common graph [3]. In complex networks, different individuals in the actual system are usually regarded as nodes, and the relationship between individuals in the system is regarded as edges. Each edge can only be associated with two nodes.

However, with the development of the actual network, the number of edges and nodes is increasing, and the types of edges and nodes are also increasing, and the network structure is becoming more and more complex. Therefore, complex networks can not fully describe the characteristics of complex systems, and in real life, complex networks can not truly reflect the characteristics of the real world. For example, in a scientific collaboration network, one edge can only describe the cooperation between two authors, but we do not know whether two or more authors are co authors of the same article; in a food network, a side can only indicate that there are common prey among species, but it does not know the composition of the entire species group with common prey; in the protein network, one edge can only describe the cooperation between two authors It can only indicate that there is the same protein between two complexes, but we can not know any other information about the protein [4]. References [5] and [6] discussed the system composed of three kinds of nodes: user, resource and tag recommendation, which are difficult to describe by complex network. To solve this problem, a natural way to represent these systems is to use the general form of graph hypergraph [7]. We can use nodes to represent individuals, and edges to represent some common characteristics of these individuals (nodes). In this way, the graph can express more information of the network. In 1970, Berge [8] proposed the basic concepts and properties of hypergraph theory. In hypergraph, a hyperedge can contain more than two nodes, which retains the original characteristics of graph. Hypernetwork based on hypergraph theory can effectively reveal the interaction between various nodes and super edges, and can effectively describe these real systems.

In recent years, with the in-depth study of hypergraph theory, hypernetwork based on hypergraph structure has developed rapidly. Researchers have constructed different hypernetwork evolution models and analyzed their characteristics. Reference [9] constructed a hypernetwork dynamic evolution model based on the hypergraph structure, the hypernetwork model added several new nodes each time, these new nodes were combined with an old node in the original network to generate a new hyperedge. Reference [10] constructed another hypernetwork dynamic evolution model, in which only one new node was added each time, and the new node was combined with several old nodes in the original network to generate a new hyperedge. In reference [11], a unified evolution model of hypernetworks and complex networks were constructed, and the evolution mechanism and topological properties of scale-free characteristics of hypernetworks were studied. In reference [12], a new hypernetwork evolution model was proposed, in which not only new nodes were added, but also the disappearance of old nodes and old hyperedges. In reference [13], the construction method of the non-uniform hypernetwork model was given, and analyzed the evolution and topological properties of the nonuniform hypernetwork model. Reference [14] focused on selfsimilar hypernetwork and random hypernetwork from the perspective of incidence matrix, and gave some properties of hypernetwork construction methods based on matrix operations. Reference [15] analyzed the mechanism of supply chain evolution based on the established hypernetwork evolution model. Reference [16] constructed a hypernetwork evolution model for scientific research cooperation based on the cooperation method of scientific research authors, and found that the author's hyperdegree distribution conforms to a power-law distribution. Reference [17] researched from the perspective of Tang poetry rhymes and vowels, constructed a Tang poetry hypernetwork model, and found that Tang poetry hypernetwork is scale-free and has high aggregation characteristics and heterogeneity. Reference [18] constructed a hypernetwork model of protein complexes, and obtained methods to identify key proteins by analyzing the characteristics of the model. Reference [19] proposed a construction method of ER random hypernetwork model, and on this basis, used different inter-layer connection methods to construct three two-layer hypernetwork models, and found the node hyperdegree distributions of these three two-layer hypernetwork models all present a bimodal distribution. Reference [20] constructed a two-layer network rumor propagation model based on Markov chain, and proposed a control strategy. Reference [21] presented the all-terminal reliability of the hypernetwork with edge failure based on the topology of the hypernetwork, and proposed two basic methods for calculating the reliability of the super network.

2 Related Work

2.1 Definition of Hypernetwork

So far, there are two definitions of hypernetwork: hypernetwork based on network and hypernetwork based on hypergraph theory. Among them, the super network based on network [22, 23] refers to the network with complex connection mode and large scale, or the multi-layer large-scale network with another network nested in one network. The network composed of this kind of network was first put forward by Nagurney, an American scientist. She called the network higher than and higher than the existing network as a supernetwork. This kind of supernetwork has the characteristics of multi-layer, multi-level, multi-attribute or multi criteria.

Hypernetwork based on hypergraph refers to the hypernetwork that can be represented by hypergraph. This kind of hypernetwork can be transformed into hypergraph for research, and some definition properties of hypergraph are used to analyze and study hypernetwork to solve practical problems in life [24]. The definition of hypergraph in mathematics is [8]: let $V = \{v_1, v_2, \cdots, v_n\}$ be a finite set. If $E_i \neq \emptyset (i = 1, 2, \cdots, e)$ and $\bigcup_{i=1}^{e} E_i = V$, the binary relation $H = (V, E)$ is called hypergraph. The elements $v_1, v_2, \cdots, v_n$ in the V are called nodes of hypergraph, and the number of nodes is expressed by $|V|$; $E = \{E_1, E_2, \cdots, E_e\}$ is the set of all hyperedges in hypergraph, and the number of hyperedges is expressed by $|E|$. The hyperdegree of a node is defined as the number of hyperedge containing the node v_i, which is recorded as $d_H(v_i)$. The degree of a hyperedge E_i is defined as the number of other hyperedges adjacent to the hyperedge, that is, the number of hyperedges with common nodes with the hyperedge. If two nodes belong to the same hyperedge, the two nodes are adjacent; if the intersection of two hyperedges is not empty, the two hyperedges are adjacent. If the sum is finite, it is called a finite hypergraph. If the number of nodes in each hyperedge is equal, it is called uniform hypergraph. Definition of incidence matrix of hypergraph [25]: the incidence matrix B of hypergraph $H = (V, E)$ is a matrix of $|V| \times |E|$, if the node v_i is in the hyperedge e_i, $b_{ij} = 1$; otherwise $b_{ij} = 0$. The node hyperdegree distribution of hypergraph refers to the probability distribution or frequency distribution of node hyperdegree in hypergraph.

2.2   Construction Method of Hypernetwork Model

(1) Each time an old node is selected to generate a new hyperedge [9].
   In the beginning, there are only a few m_0 nodes and a hyperedge containing these nodes. In each time, m_1 new node are added to the network, which generates a new hyperedge with an old node. The probability of the old node is chosen as the ratio of the node's exceeding degree to the sum of all nodes' exceeding degree in the original hypernetwork, that is to satisfy the following (1) (Fig. 1).

$$\Pi_i = \frac{d_H(i)}{\sum\limits_{j} d_H(j)} \tag{1}$$

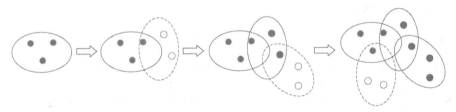

Fig. 1. The evolution process of hypernetwork model in which an old node is selected to generate a new super edge each time.

(2) Each time a new node is added to generates a new hyperedge [10].
   In the beginning, there are only a few m_0 nodes and a hyperedge containing these nodes. In each time, a new node is added to the network, which generates a new hyperedge with the existing $m_2(m_2 \leq m_0)$ old nodes in the network. The probability of the old node is chosen as the ratio of the node's hyperdegree to the sum of all nodes' hyperdegree in the original hypernetwork, that is to satisfy the following (1) (Fig. 2).

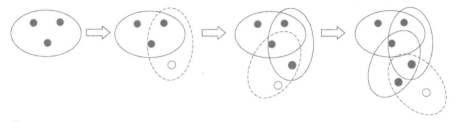

Fig. 2. The evolution process of hypernetwork model in which a new super edge is generated by adding a new node and an old node each time.

(3) Unified evolution model of hypernetworks

In reference [11], a unified uniform hypernetwork model is established. In the beginning, there are fewer m_0 nodes in the network and a hyperedge containing the number of these nodes. When m_1 new node is added to the network, these m_1 node and the existing m_2 old nodes in the network generate a new hyperedge, which symbioses into m hyperedges and doesn't appear duplicate edges. The probability of the old node is chosen as the ratio of the node's exceeding degree to the sum of all nodes' exceeding degree in the original hypernetwork, that is to satisfy the following (1) (Fig. 3).

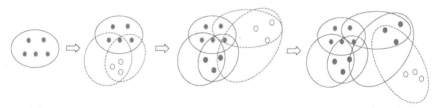

Fig. 3. Evolution process of hypernetwork unified evolution model (Each time three new nodes are added and two old nodes are selected to form two new hyperedges)

2.3 Logistic Regression Model

In 1938, Verhulst pearl proposed the logistic model. In his opinion, if a given species on the earth were placed in a specific environment, the population would be affected by the environment and could not grow indefinitely. There was an upper limit to the growth value. When the population growth is close to or close to the upper limit of growth, its growth rate will gradually decrease, and the growth rate will be slower than that at the beginning stage, and even reach a stable level. Therefore, it is also called the blocking growth model. Logistic model is usually used to simulate the population growth curve. At the beginning, human growth presents exponential explosive growth; with population growth, due to the impact of environment, resources and death, the speed of population growth will slow down; finally, the total population reaches a saturation value and tends to be stable, so this curve is also called "exponential growth" Sigmoid curve (S-shaped curve) is shown in Fig. 4.

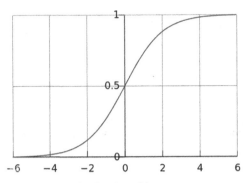

Fig. 4. Sigmoid curve

When we first studied the population forecast, the exponential model was the simplest growth model. We assume that the total population at any time t is $N(t)$ and $N(t)$ will be regarded as a continuous and differentiable function. Take the population of the initial time ($t = 0$) as N_0. The growth rate of population is defined as a constant r, that is, the increment per unit time is $N(t)$ the product of r and $N(t)$. When we consider the increase in population over time t, there are

$$N(t + \Delta t) - N(t) = r \cdot N(t) \cdot \Delta t, \tag{2}$$

Let $\Delta t \to 0$, then we can get the following differential equation

$$\begin{cases} \frac{dN(t)}{dt} = r \cdot N(t), \\ N(t)|_{t=0} = N_0. \end{cases} \tag{3}$$

In fact, this model is in good agreement with the European population growth before the 19th century. As a short-term model, it can achieve good results, but in the long run, population growth in any region can't be unlimited. Because the supply of land, water and other natural resources and the bearing capacity of the environment are limited, when the population increases to a certain number, the population growth will slow down, and the growth rate will become smaller. Therefore, the following improved model and logistic regression model are introduced.

The blocking phenomenon of population growth is mainly the change of its growth rate r, that is, the growth rate r decreases with the increase of population. We may as well express the growth rate r of population $N(t)$ as a function of the number of population $r(N_t)$, which is obviously a minus function. Therefore, (3) can be written as

$$\begin{cases} \frac{dN(t)}{dt} = r(N_t) \cdot N(t), \\ N(t)|_{t=0} = N_0. \end{cases} \tag{4}$$

Let $r(N_t)$ be a linear function of $N(t)$, i. e

$$r(N) = r_0 - s \cdot N \ (r_0 > 0, \ s > 0). \tag{5}$$

In this case, r_0 represents the growth rate when the population is small, that is, the inherent growth rate without other reasons. Suppose the maximum population capacity is N_m, which is the maximum capacity under the influence of the earth's resources and natural environment. It can be seen that $N_t = N_m$, the growth rate was zero, that is, $r(N_m) = r_0 - s \cdot N_m = 0$, so (5) becomes the following form

$$r(N_t) = r_0 \left(1 - \frac{N(t)}{N_m} \right). \tag{6}$$

It can be obtained from (4) and (6)

$$\begin{cases} \frac{dN(t)}{dt} = r_0 \cdot N(t) \left(1 - \frac{N(t)}{N_m} \right), \\ N(t)|_{t=0} = N_0. \end{cases} \tag{7}$$

(7) is the logistic regression function.

The logistic regression (LR) model is still widely used in industry, and it is the most important basic model. It is mostly used to solve classification problems in machine learning. As a growth retardation model, logistic regression model has been applied in many aspects. In this chapter, the logistic regression model is applied to the node connection in the process of super network model construction, and the characteristics of the model are analyzed.

3   Construction Method of Hypernetwork Based on Logistic Regression Model

Most of the existing hypernetwork models are based on growth and preferential connection. In the existing hypernetwork model construction, based on an initial network, each time step adds n new nodes and forms a new hyperedge with m old nodes, and finally forms a hypernetwork of network scale N, which is generally recorded as $G(G_0, m, n, N)$. In the above construction method, the new node and the old node are connected by priority connection method. However, the hypernetwork model obtained by this construction method is quite different from the actual network.

In view of the problems in the construction of the hypernetwork model, this paper improves the construction of the hypernetwork model, and obtains a hypernetwork model $G'(G_0, m, n, M, N)$ based on logical regression, where G_0 is the initial network, m is the number of old nodes selected, n is the number of new nodes, M is the maximum number of nodes in the hyperedge, N is the final network scale. There are mainly three aspects of improvement:

(1) Set the maximum capacity M of each hyperedge, that is, each hyperedge has M node at most;

(2) In the process of constructing the model, the priority connection method is replaced by the logistic regression function to deal with the connection between the new node and the old node in the network, that is, the probability of the old node being selected is

$$\Pi_i = \frac{d_H(i) \cdot (M - d_H(i))}{\sum_j d_H(j) \cdot (M - d_H(j))}. \tag{8}$$

(3) The number of old nodes is replaced by a sublinear growth function f instead of a constant C, and the number of old nodes can't exceed the growth rate of the total number of nodes in the network. Let f be the sublinear growth function of network scale $|V|$ as $f = \sqrt{|V|}$, and f can also be other types of sub linear growth function, where V is dynamic.

Table 1. Comparison between common hypernetwork model and logistic regression hypernetwork model.

Comparison standard	Network model	
	Common hypernetwork models	Logistic regression hypernetwork model
Initial network	G_0	
Number of old nodes selected	m	f
Probability of old node being selected	$\Pi_i \propto d_H(i)$	$\Pi_i \propto d_H(i) \cdot (M - d_H(i))$
Maximum number of nodes in hyperedge	∞	M
Number of new nodes	n	
Model description	$G(G_0, m, n, N)$	$G'(G_0, m, n, M, N)$

Table 1 shows the comparison between the common hypernetwork model and the logistic regression hypernetwork model.

Suppose the total size of the network is N, the capacity of each hyperedge is M, the number of new nodes added each time is n, then the number of old nodes selected each time is $\sqrt{|V|}$, the maximum number of selected nodes added each time is $\sqrt{N}$, then $\sqrt{|V|} \leq \sqrt{|N|}$, such that $n + \sqrt{G_0} \leq \sqrt{|V|} + n \leq M$. So, in this logistic regression method, When $\sqrt{|V|} + n > M$, the number of old nodes selected each time is $M - n$.

The core pseudo code of hypernetwork model based on logistic regression is as follows:

Algorithom 1. Construction method of hypernetwork model based on logistic regression

Input:	Number of network nodes before growth: m_0 ,
	Number of nodes added each time: n ,
	The maximum number of nodes beyond the in ner limit of the edge: M ,
	Network scale: N
Output:	Adjacency matrix based on logistic regression hypernetwork

1 for $i \leftarrow 1$; $i \leq m_0$; $i++$ do
2     $A(i,1) \leftarrow 1$
3 end for
4 Calculate the cumulative probability: pp
5 $S \leftarrow size(A,1)$
6 $p \leftarrow zeros(1,S)$
7 $q \leftarrow zeros(1,S)$
8 if $length(find(A == 1)) == 0$ then
9     $p(:) \leftarrow 1/S$
10 else
11 for $i \leftarrow 1$; $i < S$; $i++$ do
12     $d \leftarrow length(find(A(i,:) == 1))$
13     $q(i) \leftarrow d \cdot (M-d)$
14 end for
15 $ss \leftarrow sum(q,2)$
16 for $i \leftarrow 1$; $i < S$; $i++$ do
17     $p(i) \leftarrow q(i)/S$
18 end for
19 end if
20 $pp \leftarrow cumsum(p)$
21 Number of old nodes selected each time: m_1
22 $m_1 \leftarrow round(\sqrt{S})$
23 $e \leftarrow 1$
24 $x \leftarrow 1$
25 for $k \leftarrow m_0 + 1$; $k \leq N$; $k \leftarrow k+n$ do
26 if $k+n > N$ then
27     $n \leftarrow N-k+1$
28 end if
29 if $n < M$ && $m_1 > M$ then
30     $n \leftarrow n$
31     $m \leftarrow M-n$
32 else if $n > M$ && $m_1 < M$
33     $n \leftarrow M - m_1$
34     $m \leftarrow m_1$

```
35 else if n == M & &m₁ == M
36      n ← round(n/2)
37      m ← round(m₁/2)
38 else if n > M & &m₁ > M
39      while n + m₁ > M do
40          n ← round(n/2)
41          m ← round(m₁/2)
42      end while
43 else
44      n ← n
45      m ← m₁
46 end if
47 x++
48 for i ← 1; i < m; i++ do
49      random_data ← rand(1,1)
50      aa ← find(pp ≥ random_data)
51      jj ← aa(1)
52      A(jj,e) ← 1
53 end for
54 for j ← 1; j < n; j++ do
55      A(k + (j − 1), e) ← 1
56 end for
57 e++
58 end for
```

4 Simulation Experiment and Result Analysis

Please note that the first paragraph of a section or subsection is not indented. The first paragraphs that follows a table, figure, equation etc. does not have an indent, either.

A hypernetwork is defined as $G'(G_0, m, n, M, N)$, where G_0 represents the initial number of nodes in the network, m represents the number of old nodes in the randomly selected network, n represents the number of new nodes in the network, M represents the maximum number of nodes that each hyperedge can accommodate, and N represents the final scale of the network, that is, the final number of network nodes.

In the process of super network construction based on logistic regression, the experiment can repeatedly select old nodes, and the number of old nodes selected each time is not limited. In order to make the simulation result more stable, the experimental result is the average value repeated 50 times.

Figures 5, 6, 7 and 8 show the hyperdegree distribution of hypernetwork model based on logistic regression in double logarithmic coordinates. Figure 5 shows the hyperdegree distribution under the change of the number of old nodes m while keeping $G_0 = 50$, $n = 30$, $M = 50$, $N = 5000$ constant; Fig. 6 shows the hyperdegree distribution under the change of the number of new nodes n while keeping $G_0 = 50$, $m = \sqrt{V}$, $M = 50$, $N = 5000$ constant; Fig. 7 shows the hyperdegree distribution under the change of the number of Maximum capacity of hyperedge while keeping $G_0 = 50$, $n = 30$, $m = \sqrt{V}$, $N = 5000$ constant; Fig. 8 shows the hyperdegree distribution under the change of the number of the final scale of network formation while keeping $G_0 = 50$, $n = 30$, $m = \sqrt{V}$, $M = 50$ constant.

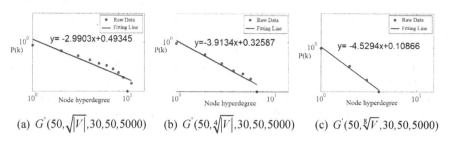

(a) $G'(50,\sqrt{|V|},30,50,5000)$ (b) $G'(50,\sqrt[4]{|V|},30,50,5000)$ (c) $G'(50,\sqrt[8]{V},30,50,5000)$

Fig. 5. Impact analysis of old node selection (m)

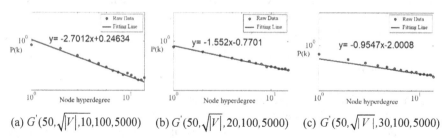

(a) $G'(50,\sqrt{|V|},10,100,5000)$ (b) $G'(50,\sqrt{|V|},20,100,5000)$ (c) $G'(50,\sqrt{|V|},30,100,5000)$

Fig. 6. Impact analysis of old node selection

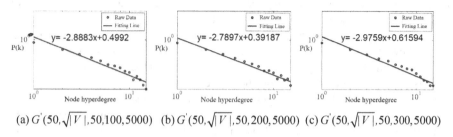

(a) $G'(50,\sqrt{|V|},50,100,5000)$ (b) $G'(50,\sqrt{|V|},50,200,5000)$ (c) $G'(50,\sqrt{|V|},50,300,5000)$

Fig. 7. Impact analysis of old node selection

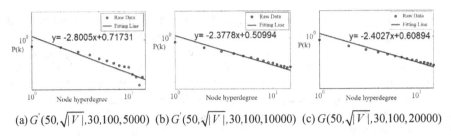

(a) $G'(50,\sqrt{|V|},30,100,5000)$ (b) $G'(50,\sqrt{|V|},30,100,10000)$ (c) $G(50,\sqrt{|V|},30,100,20000)$

Fig. 8. Impact analysis of old node selection

From Figs. 5, 6, 7 and 8, it can be observed that when the hypernetwork model based on logical regression is constructed, the straight line shows a downward trend, showing that there are fewer high value nodes and more low value nodes, showing power-law distribution and obvious scale-free characteristics.

Through comparative experiments, it is found that the change law of the hyperdegree power-law index is found by adjusting the parameters. In Fig. 5, when other parameters remain unchanged, only the number of old nodes in the hypernetwork is changed, the hyperdegree power-law index of the hypernetwork decreases with the decrease of the number of old nodes; in Fig. 6, when only the number of new nodes in the hypernetwork is changed, the hyperdegree power-law index of the hypernetwork increases with the increase of the number of new nodes; in Fig. 7, when the capacity of each hyperedge of the network is changed, the hyperdegree power-law index of the hypernetwork doesn't change much; in Fig. 8, when only the network scale is changed, the hyperdegree power-law index of the hypernetwork changes little.

5 Conclusion

Based on some problems in the process of hypernetwork model construction, this paper constructs a hypernetwork model based on logical regression, and improves the original hypernetwork model in three aspects: (1) setting the maximum capacity of the hyperedge, setting the maximum number of nodes in the hyperedge of the hypernetwork, so as not to let it grow unrestricted; (2) Set the number of connections of new nodes to show a sub linear growth; (3) the logical regression model is used to deal with the connection between nodes instead of priority selection. The experimental results show that: the hypernetwork model is power-law and scale-free. Further analysis of the experimental results shows that, the slope of the fitting line in the figure is directly proportional to when there is a change; only when there is a change, the slope of the fitting line in the figure is directly proportional to; only when there is a change, the slope rate of the fitting line in the figure only fluctuates slightly, which has little impact on the results; only if the slope of the fitting line in the figure changes When there is a change, the slope of the fitting line in the graph fluctuates only slightly, which has little influence on the results. It shows that the maximum number of nodes in each hyperedge and the final number of network nodes have little influence on the construction of hypernetwork model. By changing some parameters in the process of constructing hypernetwork model, we can get a hypernetwork model more in line with the expectation.

Acknowledgment. This work is partially supported by the National Natural Science Foundation of China under Grant No. 11661069, No. 61663041, No. 61763041, the Natural Science Foundation of Qinghai Province of China under Grant No. 2020-GX-112, and the Youth Natural Science Foundation of Qinghai Normal University under Grant No. 2020QZR007, the Chun Hui Project from the Ministry of Education of China under Grant No. Z2016101.

References

1. Watts, D.J., Strogatz, S.H.: Collective dynamics of 'small-world' networks. Nature **393**(6684), 440–442 (1998)
2. Barabási, A.L., Albert, R.: Emergence of scaling in random networks. Science **286**(5439), 509–512 (1999)
3. Sun, X.Q., Si, S.K.: Complex Network Algorithms and Applications. National Defense Industry Press, Beijing (2015)
4. Wang, Z.P., Wang, Z.T.: Theory and Application of Hypernetwork. Science Press, Beijing (2008)
5. Lü, L.Y., Medo, M., Yeung, C.H., Zhang, Y.C., Zhang, Z.K., Zhou, T.: Recommender systems. Phys. Rep. **519**(1), 1–49 (2012)
6. Ghoshal, G., Zlatić, V., Caldarelli, G., Newman, M.E.J.: Random hypergraphs and their applications. Phys. Rev. E (Statistical, Nonlinear, and Soft Matter Physics) **79**(6), 066118 (2009)
7. Berge, C.: Hypergraphs: Combinatorics of Finite Sets, 3rd edn. North-Holland, Amsterdam (1989)
8. Berge, C.: Graphs and Hypergraphs, 2nd edn. Elsevier, New York (1973)
9. Wang, J.W., Rong, L.L., Deng, Q.H., Zhang, J.Y.: Evolving hypernetwork model. Eur. Phys. J. B **77**(4), 493–498 (2010)
10. Hu, F., Zhao, H.X., Ma, X.J.: An evolving hypernetwork model and its properties. Sci. Sin. Phys. Mech. Astron. **43**(1), 16–22 (2013)
11. Guo, J.L., Zhu, X.Y.: Emergence of scaling in hypernetworks. Acta Phys. Sin. **63**(9), 090207 (2014)
12. Zhou, Z.D., Jin, Z., Song, H.T.: Emergence of scaling in evolving hypernetworks. Phys. A **546**, 123764 (2020)
13. Guo, J.L.: Emergence of scaling in non-uniform hypernetworks—does "the rich get richer" lead to a power-law distribution? Acta Phys. Sin. **63**(20), 208901 (2014)
14. Liu, S.J., Li, T.R., Hong, X.J., Wang, H.J., Zhu, J.: Hypernetwork model and its properties. J. Front. Comput. Sci. Technol. **11**(2), 194–211 (2017)
15. Suo, Q., Guo, J.L., Sun, S.W., Liu, H.: Exploring the evolutionary mechanism of complex supply chain systems using evolving hypergraphs. Phys. A **489**, 141–148 (2018)
16. Hu, F., Zhao, H.X., He, J.B., Li, F.X., Li, S.L., Zhang, Z.K.: An evolving model for hypergraph-structure-based scientific collaboration networks. Acta Phys. Sin. **62**(19), 539–546 (2013)
17. Li, M.N., Guo, J.L., Bian, W., Chang, N.G., Xiao, X., Lu, R.M.: Tang poetry from the perspective of network. Complex Syst. Complex. Sci. **14**(4), 66–71 (2017)
18. Hu, F., Liu, M., Zhao, J., Lei, L.: Analysis and application of the topological properties of protein complex hypernetworks. Complex Syst. Complex. Sci. **15**(4), 31–38 (2018)
19. Lu, W., Zhao, H.X., Meng, L., Hu, F.: Double-layer hypernetwork model with bimodal peaks characteristics. Acta Phys. Sin. (2020). https://doi.org/10.7498/aps.69.20201065
20. Yang, X.Y., Wu, Y.H., Zhang, J.J.: Analysis of rumor spreading with a temporal control strategy in multiplex networks. J. Univ. Electron. Sci. Technol. China **49**(4), 511–518 (2020)

21. Zhang, K., Zhao, H.X., Ye, Z.L., Zhu, Y.: Analysis for all-terminal reliability of hypernetworks. Appl. Res. Comput. **37**(02), 559–563 (2020)
22. Denning, P.J.: The science of computing: what is computer science. Am. Sci. **73**(1), 16–19 (1985)
23. Nagurney, A., Dong, J.: Supernetworks: Decision-Making for the Information Age. Edward Elgar Publishing, Cheotenham (2002)
24. Estrada, E., Rodríguez, V., Juan, A.: Subgraph centrality in complex network. Phys. Rev. E **71**(5), 056103 (2005)
25. Hu, F.: Research on Structure, Modeling and Application of Complex Hypernetworks. Shaanxi Normal University (2014)

Multi Dimensional Evaluation of Middle School Students' Physical and Mental Quality and Intelligent Recommendation of Exercise Programs Based on Big Data Analysis

Xinyue Li, Xin Gou, and Wu Chen(✉)

School of Software, College of Computer and Information Science,
Southwest University, Chongqing, China
chenwu@swu.edu.cn

Abstract. Taking into account that the traditional physical and mental quality education in schools is difficult to meet the comprehensive analysis and evaluation of the physical and psychological quality of middle school students, improve the enthusiasm of students to exercise physical and mental quality, let teachers and parents understand the physical and mental quality of students in real time, and provide students with physical and mental quality exercises and physical examinations. Personalized guidance for the new needs of middle school students' physical and mental quality education. We use big data analysis technology to focus on the evaluation method of middle school students' physical and mental quality and personalized exercise program recommendation method, and proposes a collaborative filtering recommendation framework based on graph neural network GNNCF (graph neural network based collaborative filtering), using Embedding technology and graph convolutional neural network to mine the attributes and interactive relationship features in the data, and then through the fusion of feature vector expressions to achieve personalized exercise program recommendations. The design and implementation of a mobile terminal-based monitoring and evaluation system for the physical and mental qualities of middle school students based on mobile terminals, "Qing Yue Circle", which can provide services for students, teachers and parents respectively, has verified that it can meet the above requirements through a fixed-point test of Qing Yue Circle in colleges and universities New demand for physical and mental quality education for middle school students.

Keywords: Big data analysis · Physical and mental quality evaluation · Physical examination · Middle school students · Personalized exercise program recommendation

© Springer Nature Singapore Pte Ltd. 2021
H. Mei et al. (Eds.): BigData 2020, CCIS 1320, pp. 209–225, 2021.
https://doi.org/10.1007/978-981-16-0705-9_15

1   Introduction

With the rapid development of society, the physical and mental health education of middle school students has arisen common concern and serious attention of families, schools and society. Moreover, our country has incorporated it into the formal school education. The traditional school education is based on the regular physical examination to understand the students' physical quality, and carry out the health education of middle school students through the unified teaching method. Although it can evaluate students' physical health and improve their physical quality to a certain extent, the traditional teaching method is still difficult to meet the needs of comprehensively analyzing and evaluating the physical and psychological quality of middle school students. It is inconvenient for teachers and parents to understand students' physical and mental quality in real time and provide personalized guidance for students' physical and mental exercise and physical examination.

For example, school physical and mental health education focuses on training and strengthening students' physical quality, but ignores the evaluation and exercise of students' psychological quality. In general, the time interval of school physical education examination is long, and these tests can only obtain the physical condition of students at a specific time point. However, it is difficult to fully grasp the changes of students' physical and mental quality in a specific period of time. Moreover, the school only regard the national standard of "students' Physical Health" issued by the Ministry of education as the standard of students' physical fitness, which ignores the analysis of the overall physical and mental quality of students. In addition, a large amount of physical health test data will be produced in the physical education of middle school students. But these data are often not effectively used to improve the enthusiasm of students to exercise physical and mental quality and provide personalized guidance for students' physical and mental quality exercise. At present, the participation of parents in school physical education is very low, which ignores the importance of parents' mastering students' physical and mental quality. Therefore, in order to comprehensively evaluate the physical and mental quality of middle school students, enhance the initiative of students to exercise physical and mental quality, give full play to teachers and parents' positive role in strengthening students' physical and mental quality, and provide personalized guidance for students' physical and mental exercise and physical examination, we need to research new solutions to assist the traditional physical and mental quality teaching in schools.

In recent years, with the development of big data technology, big data analysis method has been widely used in social network [18], e-commerce [12], search engine, health care [14], education [3], transportation, financial banking, mobile data analysis, image processing and other different fields. It can support people to excavate the value from a large amount of data to support people to make more intelligent decisions, and finally make effective countermeasures to achieve business objectives [4]. Moreover, researchers in our country have combined big data technology with education, hoping to improve the accuracy and credibility of students' comprehensive quality evaluation based on big data technology so as to promote the development of students' comprehensive quality [6,9]. Some researchers in China have also applied big data technology to students' physical health test [22,23], but their current focus is still on students' physical quality which is lack of evaluation of students' psychological quality, interaction with teachers and parents and intelligent recommendation of exercise programs for middle school students' physical examination. In particular, the lack of support for students, parents and teachers at the same time to assist students in physical and mental fitness exercise software system.

Furthermore, with the development of network representation learning, graph-based models have attracted more and more attention. When recommending an exercise program, the interaction can be modeled as a bipartite graph, where the user and the item are two independent sets of nodes, and the edges represent the interaction between the user and the item. The graph-based model combined with other attributes can explore the local topological structure and global topological structure of user item graphs, and study the effective low-dimensional representation of each user and item.

The research of collaborative filtering based on graph convolutional neural network to achieve recommendation mainly includes three directions: Network Representation Learning, High-order Proximity Learning on Graphs, and Graph-based Recommendation. (1) Network representation learning techniques aim to learn low-dimensional latent representations of nodes in a net-work. Low-dimensional vector representation can effectively maintain the local and global topological structure and node characteristics of the graph. Models based on random walking [7,15,19] were first proposed to learn this representation. Graph neural network [8,13,21] tries to adopt neural network method for graph structure data, which has developed rapidly in recent years. Graph Convolutional Networks (GCNs) [10] try to learn potential node representations by defining convolution operations on the graph. (2) The high-level structure of graphs has proven to be beneficial in many situations, such as hierarchical object representation, scene understanding, link prediction, and recommendation systems [17,20]. The network motif [16] was first proposed to learn this high-order embedding through a random walk-based model. Based on GCNs, the high-order normalized Laplacian matrix was used to aggregate the information transmitted from adjacent nodes of any order [1,11]. (3) Since user-item interaction can be regarded as a bipartite graph, many researchers are committed to using network embedding methods to learn user and item representations in the recommendation system. [2] proposed a GCN-based auto-encoder framework to complete the user-item interaction matrix. NGCF [20] stacks multiple graph convolutional layers to perform high-order embedding propagation, and connects the output of each layer as the final representation of users and items.

In spite there are many kinds of software to assist the national physical and mental health exercise at home and abroad up to now [5]. For example, Gudong, Yuedong circle, intelligent basketball, Sit-Ups as well as psychological quality training software including trendy decompression, decompression bar, etc. However, the target users of these software systems are all citizens, which are not specific to middle school students and short of specific analysis for middle school students. For example, these software systems are characterized by the separation of physical health assistance software and mental health assistance software with a single focus. There is no software system that integrates physical and psychological assistance. This is not conducive to students' comprehensive understanding of their physical and mental health. And at present, the physical examination of middle school students is becoming more and more important, but there is no special software for students' physical education examination. In addition, these softwares lack of interaction with teachers and parents, which can not make teachers and parents fully understand the actual physical and mental quality of students at any time.

Therefore, we consider the feasibility of using mobile terminal to analyze the physical and mental quality of middle school students system. This evaluation system supports the comprehensive collection of the data of middle school students' physical and mental quality, the comprehensive processing of multi-source physical and mental quality data, the multi-dimensional evaluation of the physical and mental quality of middle school students, the intelligent recommendation of exercise programs for physical and mental quality exercise and physical examination of middle school students. This paper will focus on the work of this paper from three aspects: the monitoring and evaluation system architecture of middle school students' physical and mental quality based on big data analysis, the multidimensional evaluation of middle school students' physical and mental quality and the intelligent recommendation of the physical and mental fitness exercise program of middle school students.

2   The Overall Framework of the Monitoring and Evaluation System for Middle School Students' Physical and Mental Quality Based on Big Data Analysis

The overall architecture of the monitoring and evaluation system for middle school students' physical and mental quality based on big data analysis is shown in Fig. 1. The system can be divided into user layer, access layer, service layer and data layer from top to bottom.

The user layer and access layer respectively represent the users the system is facing and the front-end access devices supported by the system. As far as the user layer is concerned, the system is oriented to four types of users: middle school students, teachers, parents and system administrators. Among them, middle school students can use the system to participate in personal physical and mental quality exercise, record exercise data, obtain physical and mental quality score, and share exercise situation or personal physical and mental quality situation. Teachers and parents can use the system to know their own class students or children's actual physical and mental quality. The administrator is responsible for the maintenance of the user information and the exercise program in the system. At the same time, it can comprehensively analyze the physical and

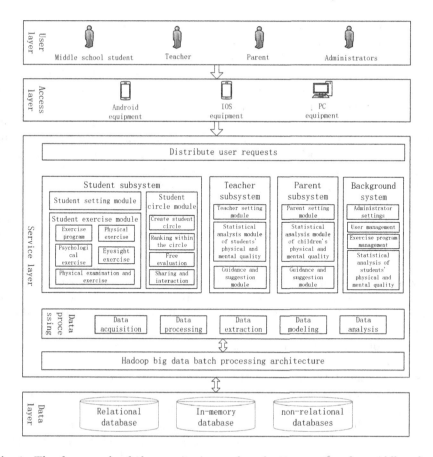

Fig. 1. The framework of the monitoring and evaluation system for middle school students' physical and mental quality based on big data analysis

mental quality of the middle school students using the system. In terms of access layer, the system designs mobile client (mobile app) for middle school students, teachers and parents, and desktop client for system administrators. Therefore, the access devices of middle school students, teachers and parents are Android or IOS devices, and administrators use personal computers to operate the system.

The service layer represents the back-end service architecture of the system. Its function is to process different requests from different types of users in the front-end user layer. The functional modules of service layer can be divided into core module and auxiliary module according to their responsibilities.

The core modules of service layer include student subsystem, teacher subsystem, parent subsystem and background system, which are used to process different types of users' requests. Among them, the basic functions of the student subsystem include: 1) recording students' personal basic information. 2) The basic data of physical and mental quality, such as figure, psychology, eyesight and physical examination results, were recorded manually. 3) Combined with the third-party physical and mental exercise auxiliary application, real-time monitoring of students' exercise (students' exercise includes learning the exercise video provided by the system, daily running and walking, sit ups, etc.). 4) This paper uses the multidimensional evaluation method of middle school students' physical and mental quality proposed in the third part of this paper to evaluate the students' physical and mental quality from multiple perspectives. 5) In this paper, the fourth part of this paper puts forward the intelligent recommendation method of physical and mental fitness exercise program for middle school students to recommend the exercise program for the students' body, psychology, eyesight and physical examination. 6) Statistics of different student circles (including peer circle, school circle, grade circle, class circle, user-defined friend circle) of students' exercise and physical and mental quality, and rank the students' physical and mental quality. The teacher subsystem supports teachers to input personal basic information and class information, and view the status quo and changes of class students' physical and mental quality. It also provides guidance to students. The parent subsystem supports parents to input personal information and their children's information, check their children's physical and mental quality and their changes, and add guidance and suggestions. The background system supports the system administrator to maintain the user information of the system, manage the exercise program, and comprehensively analyze the physical and mental quality of middle school students using the system. Among them, the exercise program refers to the different exercise contents designed by the system for students' body, psychology, eyesight and physical examination. For example, it provides fitness videos for different body parts such as shoulder, back, chest, arm, abdomen, waist, buttocks, legs and so on. Moreover, administrators need to set their properties when adding exercise programs. For example, we can indicate which category of attributes an exercise program belongs to in body, psychology, eyesight and physical examination exercise. We can also set the fitness attribute which is used to represent the score range of students' physical and mental quality corresponding to the exercise program.

The auxiliary module of service layer includes: 1) distributing user request part, which is used to receive requests from different types of users in user layer and distribute requests of specific types of users to specific user subsystem. 2) This part is used to support the real-time acquisition of students' exercise data, the evaluation of students' physical and mental quality, the recommendation of exercise programs, and the statistical analysis of students' physical and mental quality. Among them, the data acquisition module can obtain real-time exercise data (such as recording heart rate, step frequency, step number, etc.) through the mobile phone used by students or the third-party equipment (such as pedometer, smart bracelet, smart watch, etc.). Data processing module is mainly for data cleaning, data missing filling, formatting processing and other operations, that is to prepare for data analysis. Data extraction, data modeling and data analysis module, according to the data analysis request of the specific user subsystem of the service layer, carries out data extraction of a specific type or range, and carries out data analysis on individual or group of students. 3) Hadoop large data batch processing architecture part, the main function of this part is based on Hadoop framework for massive data calculation, to provide support for the service layer data processing part.

The data layer contains a variety of database servers supporting the system for data access, including relational database server and non relational database server, which are used to save the information of students, teachers, parents and administrators. These databases jointly support the business processing of the service layer.

In conclusion, the monitoring and evaluation system of middle school students' physical and mental quality based on big data analysis can monitor the physical and mental exercise of students in real time. We can complete the comprehensive analysis and evaluation of middle school students' physical and mental quality through the multi-dimensional evaluation method based on big data analysis. The system can improve the enthusiasm of students' physical and mental quality by providing comparison ranking and sharing function of physical and mental quality of students circle for middle school students. Teachers and parents timely understand the actual situation of students' physical and mental quality. By using the intelligent recommendation method proposed in this paper, students' physical and mental fitness exercise program can meet the needs of personalized guidance for students' physical and mental exercise and physical examination.

3   Multidimensional Evaluation of Middle School Students' Physical and Mental Quality

In order to let middle school students and their teachers and parents understand the comparison of students' physical and mental qualities with national standards, classmates, classmates and other angles, and the changes in students' physical and mental qualities, we propose a standard for the evaluation of physical and mental qualities of middle school students, and Based on this standard, a multi-dimensional evaluation method for physical and mental qualities of middle school students based on big data analysis is proposed.

3.1   Evaluation Standards of Physical and Mental Quality of Middle School Students

We propose data standards and time standards for evaluating the physical and mental qualities of middle school students to support the multi-angle evaluation of physical and mental qualities of middle school students.

1. Data Standards for Evaluation of Physical and Mental Quality of Middle School Students

We define two data standards. On the one hand, the standards set by the state are taken as the standards, that is, the physical health and mental health standards in the "National Student Physical Health Standard" and the "Mental Health Diagnostic Test MHT Manual" are used as the data standards for the evaluation of students' physical and psychological qualities. On the other hand, it is based on the standards of the student circle, that is, by comparing the physical and mental qualities of the students and their friends in the student circle, the students' physical and mental qualities are ranked in the student circle, so that the students can understand more about their own The position in the circle promotes mutual understanding among students and stimulates their enthusiasm for exercise. We calculate the students' physical and mental fitness scores based on these two data standards. For example, the body score interval is the score interval divided according to the BMI score table in the National Student Physical Health Standard, including [0,60), [60,75), [75,85), [85,100].

2. Time Standards for Evaluation of Physical and Mental Quality of Middle School Students

In order to enhance the accuracy and credibility of the assessment of students' physical and mental qualities, and to play a guiding role in the improvement of the students' physical and mental qualities, we define the time standards for the evaluation of the students' physical and mental qualities. We define the time standard for the evaluation of students' physical and mental quality as the single evaluation standard and time evaluation standard. Among them, the single evaluation standard is to evaluate all the physical and mental quality data of the student on the day, which allows students, teachers and parents to understand the physical and mental state of the student in a timely manner; the time period evaluation standard means that all the physical and mental quality data of the student within a specific period of time Evaluation can let students, teachers and parents understand the changes of students' physical and mental qualities.

3.2   Multi-dimensional Evaluation Method of Physical and Mental Qualities of Middle School Students Based on Big Data Analysis

The basic process of the multi-dimensional evaluation method for the physical and mental qualities of middle school students based on big data analysis we proposed is shown in Fig. 2. It is used to support the analysis and analysis of the physical and mental qualities of the students in the service layer of the monitoring and evaluation system for the physical and mental qualities of students. Evaluation. When evaluating the physical and mental qualities of middle school students, according to the evaluation criteria selected by the user, you can choose from the single evaluation based on the national student physical health standard, the period evaluation based on the national student physical health standard, the single evaluation of the student physical and mental quality based on the student circle, based on The four dimensions of student body and mind quality evaluation in the student circle are evaluated in multiple dimensions.

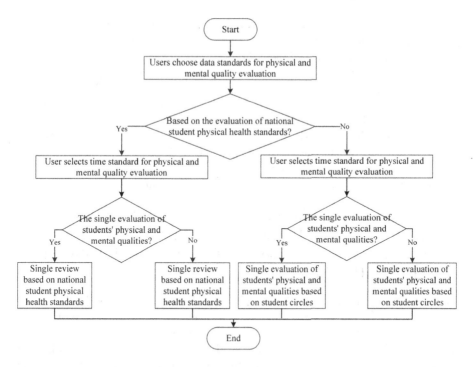

Fig. 2. The process of multidimensional evaluation of physical and mental qualities of middle school students based on big data analysis

In terms of the single evaluation dimension based on the national student's physical health standards, the system statistically counts the physical and mental quality data of the students' body, psychology, visual acuity, and physical examinations on the day. Calculate and output the gap between the scores of students' physical and mental qualities and the national standards based on the score tables given in the "National Student Physical Health Standard" and the "Mental Health Diagnostic Test MHT Manual".

In terms of the period evaluation dimension based on the national student's physical health standard, the user selects the period of time to be evaluated, the system counts the student's stature, psychology, vision and sports test scores for each day within the selected period, and output the comparison curve chart of the students' stature, psychology, vision and sports test scores and national standards in the selected period.

In terms of the single evaluation dimension of students' physical and mental qualities based on student circles, users select the student circles that need to be evaluated for their physical and mental qualities from peer circles, school circles, grade circles, class circles, and personal friend circles. Select the physical, mental, visual, and physical examination data of all students in the student circle that day, calculate and output the ranking of each student's physical and mental quality data in the selected student circle.

In terms of the evaluation dimension of the student's physical and mental quality based on the student circle, the user selects the student circle that needs to be evaluated for physical and mental quality from the peer circle, school circle, grade circle, class circle, and personal friend circle, and selects the time to be evaluated segment; the system counts the physical and mental quality data of the students and all students in the selected student circle in the selected time period for each day of the body, psychology, vision, sports exam, and calculate the ranking of the student's physical and mental quality data in the selected student circle on each day in the selected time period, and output the student user's ranking change curve of various physical and mental qualities in the selected circle in the selected time period. The quality evaluation is over.

4   Intelligent Recommendation Method of Middle School Students' Physical and Mental Quality Exercise Program Based on Graph Neural Network Collaborative Filtering

The recommendation process of the GNNCF framework proposed in this paper is shown in Fig. 3. The student subsystem at the service level in the monitoring and evaluation system of student physical and mental quality proposed in this paper recommends the appropriate physical, psychological, visual and physical examination exercise programs for middle school students respectively, so as to promote the targeted physical and mental quality exercise for middle school students. The important steps of this recommendation framework include embedding calculation, graph convolution calculation based on interaction, and recommendation based on vector fusion. The following three key steps are described in detail.

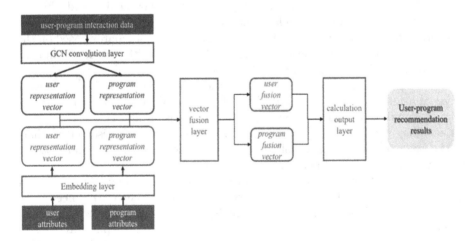

Fig. 3. Collaborative filtering recommendation process based on graph neural network

4.1 Embedding Calculation

Embedding technology is widely used in the field of deep learning by virtue of its powerful ability to integrate information features. The application of embedding layer in the recommendation system can realize the transformation from high-dimensional sparse feature vector to low-dimensional dense feature vector, so as to obtain more expressive user and item information representation vector and improve the effect of subsequent recommendation model.

The embedding layer relies on the fully connected neural network and takes the attributes of the user and the exercise program as the input respectively, to obtain the representation vectors of the user and the exercise program. Firstly, collect the user attributes containing the features of users and the program attributes containing the features of exercise programs. The user attributes include the physical and mental quality categories of users, and the program attributes include the category attributes, applicability attributes and format attributes of exercise programs (i.e., the attributes representing the display form of exercise programs, including long video, short video, text and audio). Then, one-hot encoding is performed on the obtained user attributes and program attributes, respectively, and the category-type feature attributes in the user attributes and program attributes are converted into numerical vector representations. Finally, the embedding layer of the fully connected neural network is established, and the user-related feature vector and the program-related feature vector after one-hot coding are used as the input of the embedding layer, to obtain the low-dimensional dense feature representation vector of the user and the exercise program. Take the embedding calculation of the exercise program as an example, the specific calculation is shown in equation (1):

$$R = \eta(W_1 T)$$

where, $T = \{T_1, T_2, ...T_k\}$ is a numerical vector representation after one-hot encoding, K is the number of exercise program attributes; η is the activation function of embedding calculation; R is the low-dimensional dense feature representation vector of the exercise program output after embedding calculation. Similarly, the user's low-dimensional dense feature representation vector can be calculated.

4.2 Graph Convolution Calculation Based on Interaction

The recommendation model based on graph convolutional neural network is a kind of recommendation model emerging in recent years. Relying on its powerful representation learning ability, it has excellent recommendation performance. It builds the "user-item" interaction relationship into a "node-link" relationship of the bipartite graph model, which can more fully mine the hidden associations between users and items, and recommend based on it can obtain better recommendation effect.

(1) "User-Program" Interactive Matrix Coding

According to the student user's physical and mental quality score, intelligent recommendation of physical and mental quality exercise program is carried out for student users. If the student user selects the preferred exercise program, there is a link between the student user and the exercise program to form a bipartite graph. At the same time, according to the effect of the exercise program, students score the exercise program to form a "user-program" scoring matrix. Therefore, we define student users in the system as u_i, $i \in \{1, 2, ...N_u\}$, and N_u is the total number of users; The exercise program items are defined as V_j, $j \in \{1, 2, ...N_v\}$, N_v is the total number of items. Moreover, the student users and exercise programs are constructed as nodes in the graph model, and the user ratings are constructed as links in the graph model, as shown in Fig. 4.

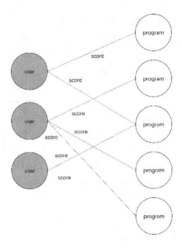

Fig. 4. Node-link graph

Encode the node vector matrix according to the graph encoder model to obtain the user representation vector and the program representation vector. The specific encoder model is shown in equation (2):

$$Y = f(X, A)$$

Where, the input parameter X is the node vector matrix of $N \times D$, N is the number of nodes in the graph, and D is the number of feature element of each node vector. Another input parameter A is the adjacency matrix of the graph. The output parameter $Y = \{y_1^T, y_2^T, ...y_N^T\}$ is the low-dimensional transformation matrix of $N \times E$, and E is the number of elements of the node representation vector in the low-dimensional space after coding transformation.

(2) Graph Convolution Representation of Node Vector

The graph convolution operation uses the classic convolution operation in CNN, and uses the weight-sharing convolution operator to operate on a subset of nodes at different positions in the graph to extract the features of the subset. Here we set the node subset S_i as a user node X_i and its first-order neighbors, then the graph convolution operation can be expressed as equation (3):

$$Z_i = C(X_i, X_i^1, ... X_i^{|N_i|})$$

Where X_i is the initial input vector of node i, $|N_i|$ is the total number of first-order neighbors of node X_i, C is the convolution operation function, and Z_i is the low dimensional space representation vector of node X_i after convolution operation.

4.3 Recommendation Based on Vector Fusion

In order to enhance the ability of node representation, the user and program vector represented by embedding and the user and program vector represented by graph convolution will be fused and expanded. Therefore, referring to the method in reference [GCMC], the node feature information processed through a layer of full connection layer and the node embedding vector obtained by the graph convolution encoder are fused, so as to obtain a new node embedding vector after feature information enhancement. The full connection model is shown in equation (4):

$$u_i = \sigma \left(W_Z Z_i + W_R R_i \right)$$

The "user-program" prediction matrix P is obtained by the inner product operation of the fused user and program vector, as shown in equation (5). And the Top-K exercise programs in the prediction vector are selected for recommendation.

$$P = u \odot v$$

5 Qing Yue Circle—Monitoring and Evaluation System of Middle School Students' Physical and Mental Quality Based on Big Data Analysis

According to the overall structure of the monitoring and evaluation system of middle school students' physical and mental quality based on big data analysis, the multidimensional evaluation method of middle school students' physical and mental quality based on big data analysis, and the intelligent recommendation method of middle school students' physical and mental quality exercise program based on big data analysis, which is called "Qing Yue Circle", is realized in this paper. The system means that the youth are happy and healthy. This paper takes the application of student end, teacher end and parent end as examples to show the basic functions of Qing Yue Circle.

As for the student end, Fig. 5 shows the student exercise part, where Fig. 5(a) shows the student's personal home page, that is, the student's personal basic information page; Fig. 5(b), 5(c), 5(d) and 5(e) respectively show the body, psychology, vision and physical examination exercise modules provided by the student end for student users, and students can choose what they are interested in or need for targeted training. Figure 6 shows the student circle part, where Fig. 6(a) shows the student circle to which the student user belongs; Fig. 6(b) shows a specific student circle described by the student user, including the name of the circle and all members; Fig. 6(c) shows the ranking of the comprehensive scores of the physical and mental qualities of all students in a specific student circle.

(a)student homepage (b)body modules (c)psychology module (d)vision module (e)physical examination modules

Fig. 5. Student exercise module display of Qing Yue Circle

Fig. 6. Display of student circle in student end of Qing Yue Circle

As for the teacher end and the parent end, Qing Yue Circle supports teachers and parents to view the physical and mental quality data of students or children at specific time points or time periods, and supports teachers and parents to provide students with guidance suggestions for physical and mental exercises. Figure 7(a) shows the teacher's view of a student's calorie consumption in a given day and the comparison analysis of the student and the average level of the class. Figure 7(b) shows parents the calorie consumption of their children during a specific period of time.

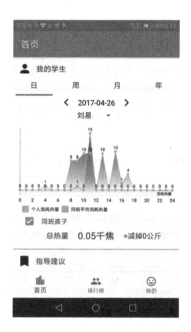 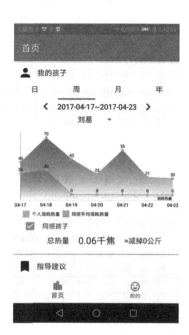

Fig. 7. Display of teacher end and parent end of Qing Yue Circle

6 Conclusion and Future Work

In order to scientifically monitor and evaluate the physical and mental health of middle school students, effectively supervise and urge them to take proper physical and mental exercises, give full play to the guiding role of teachers and parents on the physical and mental health of middle school students, and provide intelligent assistance for the physical and mental exercise and physical examination of middle school students, this paper uses big data analysis technology to design and realize a mobile terminal-based monitoring and evaluation system for the physical and mental quality of middle school students—"Qing Yue Circle". And in-depth research was conducted on the evaluation methods of physical and mental qualities of middle school students and the recommended methods of exercise programs based on graph neural network collaborative filtering. At present, the monitoring and evaluation system of middle school students' physical and mental quality designed in this paper has been used in a small range of middle schools. In this paper, we find that the performance of GNN model will be affected by the width and depth of its own network, especially in the real-time recommendation scenario of big data applications. Therefore, in the future research, we will consider and try to lightweight the GNN model to improve the scalability of GNN model. In addition, our further goal is to promote the system in a large scale. And in the process of promotion, the system is constantly improved according to the actual use.

References

1. Abu-El-Haija, S., et al.: MixHop: higher-order graph convolution architectures via sparsified neighborhood mixing (2019)
2. Berg, R.V.D., Kipf, T.N., Welling, M.: Graph convolutional matrix completion (2017)
3. Chen, J., et al.: Research on architecture of education big data analysis system. In: 2017 IEEE 2nd International Conference on Big Data Analysis (ICBDA) (2017)
4. Cheng, X.Q., Jin, X.L., Wang, Y.Z., Guo, J.F., Li, G.J.: Survey on big data system and analytic technology. J. Softw. **25**, 1889 (2014)
5. Dan-Dan, M.A., et al.: National physical monitoring and scientific fitness exercise guidance client based on IoS. Computer Technology and Development (2017)
6. Dianjun, W., Hui, J.U., Weidong, M.: Development and application of the comprehensive student quality evaluation system based on big data: innovative practice of the high school attached to Tsinghua university. China Examinations (2018)
7. Grover, A., Leskovec, J.: node2vec: scalable feature learning for networks. In: ACM SIGKDD International Conference on Knowledge Discovery and Data Mining (2016)
8. Hamilton, W.L., Ying, R., Leskovec, J.: Inductive representation learning on large graphs (2017)
9. Hong, Y., Zhu, D., Song, N., Zhou, Y.: Evaluation of students' comprehensive quality in the era of big data: Methods, values and practical guidance. China Educational Technology (2018)
10. Kipf, T.N., Welling, M.: Semi-supervised classification with graph convolutional networks (2016)
11. Lei, F.Y., Liu, X., Dai, Q.Y., Ling, W.K., Zhao, H., Liu, Y.: Hybrid low-order and higher-order graph convolutional networks (2019)
12. Li, L.: E-commerce data analysis based on big data and artificial intelligence. In: 2019 International Conference on Computer Network, Electronic and Automation (ICCNEA) (2019)
13. Li, Y., Tarlow, D., Brockschmidt, M., Zemel, R.: Gated graph sequence neural networks. Computer Science (2015)
14. Mishra, B.K., Barik, R.K., Priyadarshini, R., Panigrahi, C., Dubey, H.: An investigation into the efficacy of deep learning tools for big data analysis in health care. Int. J. Grid High Perform. Comput. **10**(3), 1–13 (2018)
15. Perozzi, B., Al-Rfou, R., Skiena, S.: DeepWalk: Online Learning of Social Representations. arXiv pre-prints arXiv:1403.6652 (2014)

16. Rossi, R.A., Ahmed, N.K., Koh, E.: Higher-order network representation learning, pp. 3–4 (2018)
17. Rossi, R.A., Rong, Z., Ahmed, N.K.: Estimation of graphlet counts in massive networks. IEEE Trans. Neural Networks Learn. Syst. 1–14 (2019)
18. Sun, Z., Dan, C.T.L., Shi, Y.: Big data analysis on social networking. In: 2019 IEEE International Conference on Big Data (Big Data) (2020)
19. Tang, J., Qu, M., Wang, M., Zhang, M., Yan, J., Mei, Q.: LINE: Large-scale information network embedding. In: International Conference on World Wide Web WWW (2015)
20. Wang, X., He, X., Wang, M., Feng, F., Chua, T.S.: Neural graph collaborative filtering. In: The 42nd International ACM SIGIR Conference (2019)
21. Xu, K., Li, C., Tian, Y., Sonobe, T., Kawarabayashi, K.I., Jegelka, S.: Representation learning on graphs with jumping knowledge networks. arXiv (2018)
22. Li, Y., Huang, Y., Lin, Q.: Design of health service system based on cloud platform (2018)
23. Yu-Fei, R., Hai-Lin, L., Hong-Fei, R.: The design and implementation of the students' physical fitness test system based on data mining. Electron. Design Eng. (2017)

Diversity-Aware Top-N Recommendation:
A Deep Reinforcement Learning Way

Tao Wang[1,2], Xiaoyu Shi[1(✉)], and Mingsheng Shang[1]

[1] Chongqing Key Laboratory of Big Data and Intelligent Computing, Chongqing Institute of Green and Intelligent Technology, Chinese Academy of Sciences, Chongqing 400714, China
xiaoyushi@cigit.ac.cn
[2] University of Chinese Academy of Sciences, Beijing 100049, China

Abstract. The increasing popularity of the recommender system deeply influences our decisions on the Internet, which is a typical continuous interaction process between the system and its users. Most previous recommender systems heavily focus on optimizing recommendation accuracy while neglecting the other important aspects of recommendation quality, such as diversity of recommendation list. In this study, we propose a novel recommendation framework to optimize the recommendation list for the Top-N task, named Collaborative Filtering-based Deep Reinforcement Learning (CFDRL), which promotes the diversity of recommendation results without sacrificing the recommendation accuracy. More specifically, to effectively capture the continuous user-item interaction for recommendations, we adopt the deep reinforcement learning (DRL) to update the recommendation strategy dynamically according to the user's real-time feedback. Meanwhile, to generate diverse and complementary items for recommendation, we design a diversity-aware reward function that can lead to maximizing reward with the trade-off between diversity and accuracy. Besides, to alleviate the disadvantage of DQN that directly picking the recommendations with the highest Q-values from the unselected items, we define a modified *ε-greedy* explore policy with jointly CF model. It firstly utilizes CF model to sort the items and divide them into two part according to the item similarity, then with a probability the agent selects from them and generates an action list with the modified *ε-greedy* explore policy. The experimental results conducted on two real-world e-commerce datasets demonstrate the effectiveness of the proposed model.

Keywords: Recommender system · Deep reinforcement learning · Collaborative filtering · Diversity-aware reward

1 Introduction

Promoted by the rapid development on the Internet of Thing (IoT) devices and advanced machine learning technologies, the last decade has witnessed a data explosion in Internet services and contents. As a result, it becomes increasingly difficult for people to retrieve their truly desired information from such a big amount of data. In this context, recommender systems (RS), as an intelligent e-commerce application, can help users

H. Mei et al. (Eds.): BigData 2020, CCIS 1320, pp. 226–241, 2021.
https://doi.org/10.1007/978-981-16-0705-9_16

search for information that may be of interest and useful by suggesting items (products, services) that best fit their needs and preferences. Recommender systems have become increasingly popular and are widely applied in a variety of fields, such as movies, music, books, points of interests, and social events. For example, Amazon [2], Microsoft [3], and JD [4] have extensively support recommendation services to improve their users' shopping experiences.

Recommender based on variant ideas and technologies can be further classified into content-based recommender, collaborative filtering (CF)-based recommender and hybrid recommender. Among them, CF-based recommender is the most popular and widely used in industrial. The basic idea is that it first calculates the similarity between target users or items, then predicts target users' preference for items based on the evaluation of users with similar interests to target users. Recently, the vigorous development of deep learning has brought new vitality to recommender systems. Dependent on the strong data representation ability, deep learning can effectively capture nonlinear and non-trivial user-item relationships, and encode more complex abstractions as high-level data representations. Besides, it also captures complex relationships in the data from a rich set of accessible data sources. Hence, it has widely applied to recommender designing for improving the recommendation accuracy.

In general, the recommender system is a typical consistent user-item interaction process. For the top-N task, the system first generates a recommendation list and sends it to its user, then user provides feedback based on the received recommendation list. After that, the system recommends a new set of items based on user feedback. Hence, the user's decision on each item could deeply affect the performance of the recommender system in real-world environments. To effectively capture such complex and nonlinear interaction between the system and its users, deep reinforcement learning (DRL) technologies have emerged to the recommender system, which utilizes to model the continuous interaction process [14]. DRL-based recommender systems have two major advantages: 1) DRL based recommender can update their recommendation strategies on-the-fly follow by the user's real-time feedback during the interactions until the system converges to the optimal strategy that generates recommendations best fitting users' dynamical preferences. 2) In DRL-based methods, the optimal strategy is made by maximizing the expected long-term cumulative reward from designed reward function; while the majority of traditional recommender systems focus on maximizing the immediate (short-term) reward of recommendations, leaving the user to focus only on the items that are currently rewarded the most, completely ignoring whether those items will yield more likely or profitable (long-term) returns in the future. Therefore, DRL-based methods are more conducive to the health of system, since they can identify items with small immediate rewards but making big contributions to the rewards for future recommendations.

DRL solutions can recommend a set of items each time. For example, DQN generates a recommendation list with the highest Q-values according to the current state [14]. However, these approaches generate recommendations based on the same state, which leads to the recommended items to be similar. In practice, a recommendation list with diverse and complementary items may receive higher rewards than recommending the homogeneous items. However, most previous recommender systems heavily focus on optimizing recommendation accuracy while neglecting the other important aspects of recommendation quality, such as diversity of recommendation list since the original intention of employing the recommender system is that not only mining some useful information to user, but also bringing profits to the sites.

Differing from existing recommendation methods, this paper proposes a novel recommendation framework, namely Collaborative Filter based Deep Reinforcement Learning (CFDRL), for improving the diversity of recommendation without sacrificing the recommendation accuracy. More specifically, for promoting the diversity of recommendation results, we design a novel diversity-aware reward function and optimize the exploration policy to generate the recommendation list correspondingly. The diversity-aware reward function representation user' feedback and evaluates the quality of the recommendation policy, considering both current and future rewards for the recommended items. To take advantage of a reward function and explore policy, we adopt the Deep Q-Learning reinforcement learning framework. For each user, the recommendations are dynamically generated when an item is given. Furthermore, in CFDRL, we adopt the CF model to generates two action parts, then with a probability Q network select from them and generate an action list with the defined explore policy.

We summarize our major contributions as follows:

- We propose a novel framework CFDRL to generate diverse and complementary items for recommendation automatically according to user's real-time feedback. Which can promote the diversity of recommendations without losing the recommendation accuracy.
- We define a diversity-aware reward function that can reflect the accuracy of recommendation and the diversity in the same time. And a more effective exploration method modified *ε-greedy* is applied, it adopts the CF model to generate the candidate items as an action pool, rather than directly using the unselected items as action pool.
- We conduct extensive experiments on two real-world datasets and validate the performance of our proposed model. Compared with the state-of-the-art solutions, the results show that our proposed recommender achieves a better performance in terms of recommendation accuracy, diversity and novelty of the recommender system.

The rest of this paper is organized as follows. Section 2 reviews the related works. Section 3 describes the proposed framework. Section 4 empirically evaluates the proposed recommendation model. Finally, Sect. 5 discusses and concludes this paper.

2   Related Work

Reinforcement learning has been extensively studied in the field of recommendation. Mahmood et al. [14] adopted the reinforcement learning technique to observe the responses of users in a conversational recommendation, intending to maximize a numerical cumulative reward function modeling the benefit that users get from each recommendation session. Sunehag et al. [15] introduced agents that successfully address sequential decision problems with high-dimensional combinatorial slate-action spaces. However, these methods put a heavy focus on the recommendation accuracy with less consideration of diversity and novelty of recommendations.

Promoting the diversity of recommendation results has received increasing research attentions. The maximal marginal relevance (MRR) model [17] was one of the pioneering works for promoting diversity in information retrieval tasks. Lathia et al. designed and evaluated a hybrid mechanism to maximize the temporal recommendation diversity in [18]. Zhao et al. [19] utilized the purchase interval information to increase the diversity of recommended items. Qin and Zhu proposed an entropy regularizer to promote recommendation diversity. This entropy regularizer was incorporated in the contextual combinatorial bandit framework to diversify the online recommendation results in [20]. Recently, a maximum-weight degree-constrained subgraph selection method was proposed to post-process the recommendations generated by collaborative filtering models, so as to increase recommendation diversity [21]. In [22], the diversified recommendation problem was formulated as a supervised learning task, and a diversified collaborative filtering model was introduced to solve the optimization problems. In [23], a fast greedy maximum a posteriori (MAP) inference algorithm for determinantal point process was proposed to improve recommendation diversity. Different from the above methods, we utilize a DRL-based method to conduct the diversity-aware recommendations, the most attractive part of DRL-based method is that the recommend result will be optimized in the continuous interaction process.

3 The Proposed Recommender

In this section, we first introduce the CFDRL framework in a high-level way. Then, the proposed diversity-aware reward function and modified ε-greedy explore strategy are given. Finally, we present the training algorithm of CFDRL.

3.1 The CFDRL Framework

We model the recommendation session as a Markov Decision Process (MDP), which includes a sequence of states, actions and rewards. Formally, MDP consists of a tuple of five elements (S, A, R, P, γ) as follows:

- **State space** S: A state $s_t = \{s_t^1, \ldots, s_t^N\} S$ is defined as the browsing history of a user, i.e., previous N items that a user browsed before time t. The items in s_t are sorted in chronological order.
- **Action space** A: an action $a_t = \{a_t^1, \ldots, a_t^K\} A$ is to recommend a list of items to user at time t on account of current state s_t, where K is the number of items the RA recommends to user each time.
- **Diversity-aware Reward Function** R: After the recommender system takes an action a_t at the state s_t, i.e., recommending a list items to a user, the user browses these items and gives her feedback, and the system receives immediate reward $R(s_t, a_t)$ according to such feedback.
- **Transition probability** P: Transition probability $P(s_{t+1}|s_t, a_t)$ is defined as the possibility if the users can turn their state from the current state s_t to a state s_{t+1}, Since $P(s_{t+1}, a_t, \ldots s_1, a_1) = P(s_{t+1}|s_t, a_t)$ in MDP [6], if a user is interested in what the system recommends according to our definition, then the next state is s_t, otherwise, the user show little interest in the recommendation, then the next state stays at s_t.

- **Discount factor** γ: when we calculate the value of future rewards, $\gamma[0,1]$ defines the discount rate. In particular, when $\gamma = 0$, the system only considers the current rewards. In other words, when $\gamma = 1$, all future rewards are taken into account.

In practice, it is not sufficient to represent items only in discrete indexes, since we do not know the relationship between the different items just through the index.

From the above description, the recommendation problems can be formally defined as follows: given the historical MDP, i.e., (S, A, R, P, γ), the goal is to find an optimal recommendation policy $\pi: S \to A$, that can maximize the cumulative rewards for the recommender system.

The optimal action-value function obeys an important identity known as the Bellman equation. This is based on the following intuition: if the optimal value $Q^*(s', a')$ of the sequence s' at the next time-step was known for all possible actions a' maximizing the expected value of $r + \gamma Q^*(s', a')$:

$$Q^*(s, a) = E[r + \gamma \max_{a'} Q^*(s', a')|s, a] \tag{1}$$

The basic idea of reinforcement learning algorithms is to estimate an iteratively updated action value function by using the Bellman equation $Q_{i+1}(s,a) = E[r + \gamma \max_{a'} Q_i(s', a')|s,a]$. This algorithm converges to obtain the optimal action value function, $Q_i \to Q^*$ as $i \to \infty$. In practice, this basic approach is impractical because the action-value function is estimated individually for each sequence without any generalization. Instead, function approximator is usually used to estimate action value functions $Q(s, a; \theta) \approx Q^*(s, a)$ in reinforcement learning communities, which is usually a linear function approximator. But sometimes nonlinear function approximation is used instead, such as neural network. We refer to a neural network function approximator with weights θ as a network. A Q-network can be well trained by adjusting the parameters θ_i at iteration i to reduce the mean-squared error in the Bellman equation, where the optimal target values $r + \gamma \max_{a'} Q^*(s', a')$ are substituted with approximate target values $y = r + \gamma \max_{a'} Q(s', a', \theta_i^-)$, using parameters θ_i^- from some previous iteration. This leads to a sequence of loss functions $L_i(\theta_i)$ that changes at each iteration i (Fig. 1).

$$L_i(\theta_i) = E_{a,r}[(E_{s'}[y|s, a] - Q(s, a; \theta_i))^2]$$
$$= E_{a,r,s'}[(y - Q(s, a; \theta_i))^2] + E_{s,a,r}[V_{s'}[y]] \tag{2}$$

Note that the target depends on the network weight; This is the opposite of the goal of supervised learning, which is defined before learning begins. The last term is the variance of the target, which does not depend on the the parameter that we are currently optimizing, so it can be ignored. By differentiating the weight of the loss function, the following gradient can be obtained:

$$\nabla_{\theta_i} L(\theta_i) = E_{a,r,s'}[(r + \max_{a'} Q(s', a'; \theta_i^-) - Q(s, a; \theta_i))\nabla_{\theta_i} Q(s, a; \theta_i)] \tag{3}$$

3.2 The Diversity-Aware Reward Function

To promoting the diversity of recommendations, we define the reward function for the sake of improving the diversity and the novelty of recommendation. As traditional deep

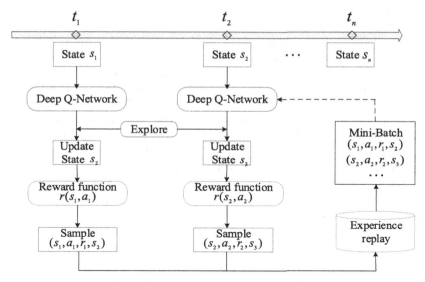

Fig. 1. Recommendation process modeling

reinforcement consider the reward a simple feedback more (i.e., usually setting it as 1/0 or score representing the feedback), the result gets the maximum long-term return based on defined reward. Therefore, the definition of reward should also be considered properly and professionally.

After the recommender system takes an action at the state st, i.e., recommending a list items to a user, the user browses these items and gives her feedback, and the system receives immediate reward R(st, at) according to the feedback.

The reward is defined as follows:

$$R = \omega_1 \sum_{n=1}^{N} \gamma^n \cdot sim(s[n], a) + \frac{\omega_2}{lg(pop(a))} + \omega_3 \varepsilon_i \qquad (4)$$

Where ω_1, ω_2 and ω_3 indicates the weight of every part, γ represent the coefficient we add on sim. ε_i indicates the the influence of uncertain factors, here we use the Gaussian noise. $s[n]$ and a mean the nth item in the state and the action which recommends by the system. The first part sim is presented in (5) and pop in second part is in (6).

$$sim(s[n], a) = \frac{num(s[n]) * num[a]}{\sqrt{|num(s[n])| * |num[a]|}} \qquad (5)$$

$$pop(a) = \frac{num[a]}{nums} \qquad (6)$$

Where $nums(s[n])$ indicates the number of user who browses item $s[n]$, and $nums$ represents the total users.

3.3   Modified ε-*greedy* Explore

The most straightforward strategies to do exploration in reinforcement learning are
ε-*greedy* [14] and UCB [1]. ε-*greedy* will randomly recommend new items with a probability of ε, while UCB picks items that have not been explored for many times (because
these items may have larger variance). It is evident that these trivial exploration techniques harm recommendation performance in a short period. Therefore, rather than doing
random exploration, we propose a modified ε-*greedy* to do the exploration. Intuitively,
as shown in Fig. 2, we sort the items according to the CF model, then we divide the item
space into two parts M and N. With the probability ρ we push the items in M and the
probability $1 - \rho$ in N to the network Q. The network Q calculates every value of the
items, and the calculated cost can not be enormous because of the limited length of M,
this meanwhile solves the problem of state space and action space explosion problems
in reinforcement learning.

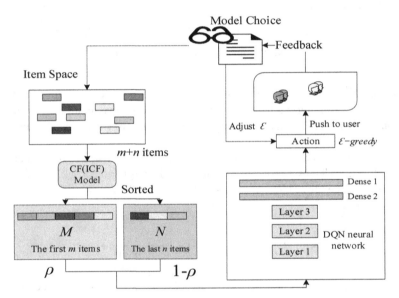

Fig. 2.   Exploration by modified ε-*greedy*

As the select policy in Q network, it learns about the greedy policy $a = \mathrm{argmax}_{a'}Q(s,$
$a'; \theta)$, while following a behavior distribution that ensures adequate exploration of the
state space. In practice, the behavior distribution is usually selected by an ε-*greedy*
strategy that we will adjust ε according the feedback which also means reward. While
the reward is over a threshold value we set, ε is to be decrease until it reach its minimum
we set. The action push to user will get feedback which we define as the diversity-aware
reward function will be used to update the original action space. The explore policy we
modify increases the update speed of network Q and improves the diversity which we
focus on most with a low decrease of accuracy.

3.4   The Proposed Algorithm

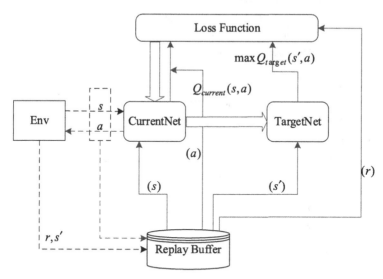

Fig. 3. Process and training

The training algorithm for proposed framework CFDRL is presented in Algorithm 1. Each iteration consists of two steps, i.e., 1) transition generating stage (lines 9–21). And 2) parameter updating stage (lines 22–30). For the transformation generation phase (line 9): given the current state s_t, the network first gives the action list $a_t = \{a_t^1,...,a_t^k\}$ according to the reward value calculation from Algorithm 1 (line 10) and updates the state to s_{t+1} (lines 12–18) following the same strategy in Algorithm 1; Finally the recommendation agent stores transitions (s_t, a_t, r_t, s_{t+1}) into the memory D (line 20), and sets $s_t = s_{t+1}$ (line 21). For parameter updating stage: the recommendation agent samples mini-batch of transitions (s, a, r, s') from D (line 23). And then the network parameters of the network are updated (lines 24–29) (Fig. 3).

In the algorithm, we introduce the widely used techniques to train our framework. For example, we utilize a technique known as experience replay (line 1, 23), and introduce separated evaluation and target networks (line 2, 24), which can help smooth the learning and avoid the divergence of parameters. For the soft target updates of target networks (lines 26), we used $\tau = 0.01$. Besides, we use the monte-carlo sampling strategy to assist the framework learning from the most important historical transitions and increase the variety of items.

Algorithm 1 Parameters Training for recommendation with CFDRL.

1: Initialize replay memory D to capacity N

2: Initialize action-value function Q with random weights θ

3: Initialize target action-value function Q' with weights θ

4: Initialize the defined boundary reward value r_0

5: **For** episode=1, M **do**

6:　　Reset the item space I

7:　　Initialize state s_0 from previous record

8:　　**for** t=1, T **do**

9:　　　　**Stage 1: Transition Generating Stage**

10:　　　　　　Select an action $a_t=\{a_t^{1},...,a_t^{K}\}$

11:　　　　　Execute action a_t and observe the reward r

12:　　　　　Set $s_{t+1}=s_t$

13:　　　　　**for** k=1, K **do**

14:　　　　　　**if** $r_t^{K}>r_0$ **then**

15:　　　　　　　Add a_t^{K} to end of s_{t+1}

16:　　　　　　　Remove the first item of s_{t+1}

17:　　　　　　**end if**

18:　　　　　**end for**

19:　　　　　Compute the overall reward r_t according Eq.(1)

20:　　　　　Store transition (s_t,a_t,r_t,s_{t+1}) in D

21:　　　　　Set $s_t=s_{t+1}$

22:　　　　**Stage 2: Parameter Updating Stage**

23:　　　　　Sample minibatch of N transitions (s,a,r,s') from D

24:　　　　　Generate a' by actor network

25:　　　　　Set $y=r+\gamma\max_{a'}Q(s',a',\theta_i^{-})$

26:　　　　　Perform a gradient descent step on $y-Q(s_t,a_t;\theta)^2$ with respect to the network parameter θ

27:　　　　　Every C steps reset $Q'=Q$

28:　　**end for**

29: **end for**

4　Experiments

In this section, we conduct a number of experiments on two well-known industrial recommender systems, i.e., Movielens and Netflix. The details of datasets are described in Table 1. All experiments are conducted on a tower server, which includes 2.1 GHz, Intel Xeon E5 CPU, 250 GB RAM, Ubuntu 18.04LTS operation system, JDK1.8, Python3.6 and Keras. We mainly focus on two questions: (1) How about our proposed reward function and explore policy performs; (2) How the proposed framework performs compared to representative baselines. We first introduce experimental settings. And then we seek answers to the above two questions. Finally, we conclude the results on the performance of the proposed framework.

Table 1. Dataset in details.

Dataset	#Users	#Items	#Ratings
Movielens	138,493	26,744	20,000,263
Netflix	480,189	98,212	100,480,507

4.1 Experimental Settings

Parameter Setting: We adopt target-Q network with experience replay. The network will be updated by using Algorithm 1 with $\tau = 0.01$, and we set the capacity of replay memory $D = 2000$, the size of minibatch $N = 64$. We use the Adam optimizer to optimize the network. The learning rate for network are 0.005 respectively. After 500 steps of exploration, the network start to train, the probability of exploration is reduced from 0.95 to 0.05, and the rate of decline is 0.995. We keep $n = 20$ in (1). For each dataset, the ratio of training, evaluation, and test set is 6:2:2. Each experiment is repeated 5 times, and the average performance is reported.

Evaluation Metrics: In order to evaluate our experimental results, we use accuracy precision@K, inter-item diversity *Interdiv* [6] and intra-item diversity as the metrics to evaluate the overall results. $R(u)$ of formula (7) represents the items in the recommendation list of the training set while $T(u)$ represents list of the training set. In formula (8) (9) (11) (12) (13), L_u and L_v represents a list of recommendations to the user. The definition of *interdiv* for the similarity between classes is defined and represented by *interdiv*. The similarity within the class are called *ILS* [6]. Finally, the popularity metric is represented by *Novelty* [12] in traditional recommender systems. The metrics are as follows:

$$Precision@K = \frac{\sum_{u \in U} |R(u) \cap T(u)|}{|U| \times K} \tag{7}$$

$$Interdiv = \frac{2}{n(n-1)} \sum_{u,v \in U, u \neq v} \frac{|L_u \cap L_v|}{|L_u|} \tag{8}$$

$$ILS(L_u) = \frac{2 * \sum_{i,j \in L_u, i \neq j} similarity(i,j)}{|L_u| * (|L_u| - 1)} \tag{9}$$

We calculate *similarity(i,j)* as (10):

$$similarity(i, j) = \frac{\Gamma(i) \cap \Gamma(j)}{\Gamma(i) \cup \Gamma(j)} \tag{10}$$

Where $\Gamma(i)$ represents the times item i show in all recommend lists.

$$ILS = \frac{1}{n} \sum_{u \in U} ILS(L_u) \tag{11}$$

$$Novelty(L_u) = \frac{\sum_{i \in L_u} P(i)}{L_u} \tag{12}$$

$$Novelty = \frac{1}{n} \sum_{n \in U} Novelty(L_u) \tag{13}$$

Where $P(i)$ indicates the average number of occurrences of an item in all items.

From (7), we know the value of Pre@K represents the accuracy of recommend by some method. So a higher result of Pre@K indicates this method is better. However, from the definition of *Interdiv* and *ILS*, we gain that the similarity of recommend results will be high if the value of result achieve large. In other words, the diversity of recommend we focus on will be bad. As a result, the smaller value of *Interdiv* and *ILS*, the better performance a method will be. The same is true for *Novelty*. We use the average popularity of recommend items to reflect novelty, it is one of the important indicators that affect user experience. And it refers to the ability to recommend non-popular items to users. The smaller value of *Novelty* will reflect the better result.

Baselines: We select the state-of-art recommendation approaches for comparison:

- **UCF**: User collaborative filtering [2] is a method of making automatic predictions about the interests of a user by collecting preference information from many users, which is based on the hypothesis that people often get the best recommendations from someone with similar tastes to themselves.
- **ICF**: Item collaborative filtering [2] is a method that automatically predicts users' interests by collecting many users' preference information based on the assumption that people often choose items that have the same characteristics and users' interests as the selected items.
- **LinUCB**: Linear Upper Confidence Bound [1] can select an arm (i.e., recommend a piece of news) according to the estimated upper confidence bound of the potential reward.
- **C²UCB**: [13] This is a contextual bandit based interactive recommendation method, which promotes the recommendation diversity by employing an entropy regularizer.
- **D²RL**: [16] encourages the diversity of recommendation results in interaction recommendations adopt a Determinantal Point Process (DPP) model to generate diverse, while relevant item recommendations through an actor-critic reinforcement learning framework.

4.2 Diversity-Aware Reward Function and Explore Analysis

As described in (1), to be fair, we carry out our experiments setting the recommend list $K = 20$ on all methods. We use Deep Q-network with consideration future reward named as "DQN", Then, by adding CF model to select action for network, this becomes "CF +

Table 2. Overall performance comparison (the best scores are highlighted.)

Dataset	Models	Pre	Interdiv	ILS	Novelty
Movielens	CF	**0.1434**	0.4736	0.4317	0.2534
	C^2UCB	0.0412	0.2535	0.2148	0.2354
	D^2RL	0.1046	0.2054	0.1956	0.1214
	DQN	0.0927	0.2314	0.2167	0.1694
	CF + DQN	0.1196	0.1434	0.1245	0.1042
	CF + DQN + ME	0.1194	**0.1268**	**0.0954**	**0.1037**
Netflix	CF	**0.1624**	0.5254	0.5142	0.3425
	C^2UCB	0.0721	0.2545	0.2013	0.2142
	D^2RL	0.1145	0.1634	0.1845	**0.1047**
	DQN	0.0847	0.1547	0.1431	0.1874
	CF + DQN	0.1234	0.1614	0.1546	0.1647
	CF + DQN + ME	0.1347	**0.1134**	**0.1254**	0.1354

We care about how our proposed model is affected by the diversity-aware reward and explore function.

DQN". "CF" means ICF. After that, we add more components to "CF + DQN". "ME" stands for the modified ε-*greedy* explore policy. So "CF + DQN + ME" represents our model CFDRL. The performance on these datasets in terms of all the metrics is shown in Table 2.

The result is shown in Table 2. "Pre" represents "Pre@K". As expected, our algorithms outperform all baseline algorithms. Our base model DQN already achieves relatively good results. This is because the dueling network structure can better model the interaction between user and item. Adding CF model, we achieve another significant improvement mainly on accuracy, as the CF model will help network select similar actions to promote accuracy. With random exploration network will reduce the similarity and improve the diversity of recommendation, which illustrates the effectiveness of DQN adding CF model. Besides, by using the modified exploration policy, our model achieves the best result in *Interdiv* and *ILS* on both datasets, and it can maintain a good result of accuracy as expected.

4.3 Performance Comparison on Movielens

First we evaluate the performance of all the methods on Movielens. Figure 4 shows how these methods perform with the variety of length of recommend list. We offer the following suggestions from Fig. 4:

- In the recommendation sessions, for the metric Pre@K, though ICF consistently outperforms all the methods, the result of CFDRL approaches a relatively good level than the other baselines. As our goal is promoting diversity without great damage on

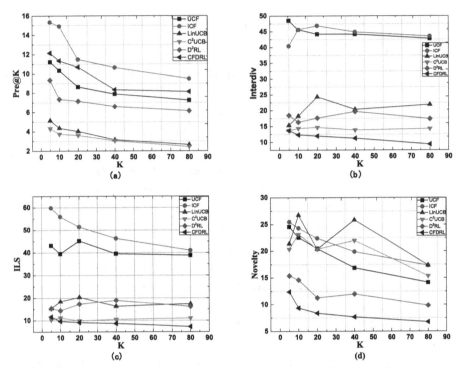

Fig. 4. Performance comparison of different algorithms on Movielens. Figure 4(a) shows the Pre@K of different methods. Figure 4(b) shows the metric *Interdiv* of different methods. Figure 4(c) shows the comparison on *ILS*. Figure 4(d) shows the performances about *Novelty*

accuracy, the performance of accuracy for CFDRL can be accepted. The result proves that our proposed method can work well in the movie recommendation scenario with a relatively high accuracy.

- For one of the most important metrics *Interdiv* that we care about, CFDRL consistently outperforms all the baselines in terms of recommendation on Movielens dataset. Moreover, from Fig. 4(c), we can also note that CFDRL also consistently generates more diverse recommendations than the baselines with the different length of recommend lists. We defines the diversity-aware reward function in (1) which includes similarity with a attenuated coefficient between recommend items and user browsing records, that makes the similarity decrease effectively. We can figure out that as the length of recommend list varies, the result of metric *Interdiv* and *ILS* grow more competitive. When $K = 80$, they come to optimal performance, and the interaction between users and items are more sufficient, which makes the advantage of CFDRL more obvious.

- The metric *Novelty* shows CFDRL performs best than all other methods in terms of different length of recommend lists. For our defined reward function is able to optimize the novelty through the interaction between user and agent.

4.4 Performance Comparison on Netflix

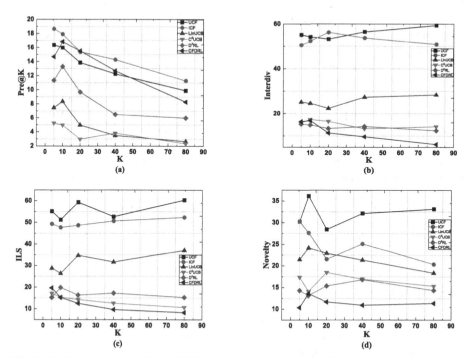

Fig. 5. Performance comparison of different algorithms on Netflix. Figure 5(a) shows the Pre@K of different methods on Netflix. Figure 5(b) shows the metric *Interdiv* comparison of different methods. Figure 5(c) shows the results on *ILS*. Figure 5(d) shows the performances about *Novelty*.

To verify the generalization ability of our proposed model, we conduct the same experiments on Netflix. All parameters and settings are exactly the same as experiments on Movielens.

The results of experiments on Netflix are shown in Fig. 5. It can be observed for the metric Pre@K, ICF outperforms all the other methods just as it on Movielens. But for *Interdiv*, CFDRL achieves the best result in most situations. D^{2}RL performs better when the recommend list is 5 and 10, that is because this method optimize accuracy as well as diversity, the result can approach a good result. Superior performance of CFDRL can also be observed for the metric of *ILS*. Both C^{2}UCB and CFDRL achieve stable performance on recommendation diversity, and CFDRL has much better result compared to C^{2}UCB in accuracy. In addition, CFDRL performs best for the metric of Novelty than other baselines in recommendation. As a result, CFDRL shows strong ability in terms of recommendation for these situations which are supposed to trade-off the accuracy and diversity. Our experiments on two different datasets achieve similar results, which illustrates the generalization ability of CFDRL is considered well and our model is able to applied to different areas for recommendation.

5   Conclusion

In this paper, we propose a novel recommendation framework, named Collaborative Filtering-based Deep Reinforcement Learning (CFDRL). Specifically, CFDRL utlizes deep reinforcement learning framework to model the interactions between users and the recommender system, and learn the optimal recommendation strategies dynamically based on user's real-time feedback. Different from previous work, we propose the diversity-aware reward function to maximize the long-term cumulative reward with trade-off the recommendation accuracy and diversity. Furthermore, we proposed a modified ε-greedy explore policy to generate the recommendations. Extensive experiments on two real world datasets have been performed to demonstrate the effectiveness of CFDRL. The future work will focus on the follow directions. First, we would like to try other reinforcement learning model like DDPG. Second, we prefer to design more efficient reward function for generating relevant yet diverse recommendation.

Acknowledgement. This research is in part by Chongqing research program of key standard technologies innovation of key industries under grant cstc2017zdcyzdyfX0076, cstc2019jscx-zdztzxX0019, in part by Youth Innovation Promotion Association CAS, No. 2017393.

References

1. Li, L., Chu, W., Langford, J., Schapire, R.E.: A contextual bandit approach to personalized news article recommendation. In: Proceedings of the 19th International Conference on World Wide Web, pp. 661–670. ACM (2010).
2. Linden, G., Smith, B., York, J.: Amazon.com recommendations: Item-to-item collaborative filtering. IEEE Internet Comput. **7**, 76–80 (2003)
3. Sunehag, P., Evans, R., Dulacarnold, G., et al.: Deep reinforcement learning with attention for slate markov decision processes with high-dimensional states and actions. arXiv: Artificial Intelligence (2015)
4. Zhao, X., Xia, L., Zhang, L., Ding, Z., Yin, D., Tang, J.: Deep reinforcement learning learning for page-wise recommendations. In: RecSys 2018, pp. 95-103 (2018)
5. Mnih, V., Kavukcuoglu, K., Silver, D., et al.: Playing atari with deep reinforcement learning. arXiv: Learning, (2013)
6. Zhang, M., Hurley, N.: Avoiding monotony: improving the diversity of recommendation lists. In: RecSys 2008 (2008)
7. Li, L., Chu, W., Langford, J., Schapire, R.E.: A contextual-bandit approach to personalized news article recommendation. In: WWW 2010 (2010)
8. Kawale, J., Bui, H.H., Kveton, B., Tran-Thanh, L., Chawla, S.: Efficient Thompson sampling for online matrix factorization recommendation. In: NIPS 2015 (2015)
9. Chen, H., et al.: Large-scale interactive recommendation with tree-structured policy gradient. In: AAAI 2019 (2019)
10. Zhao, X., Zhang, L., Ding, Z., Xia, L., Tang, J., Yin, D.: Recommendations with negative feedback via pairwise deep reinforcement learning. In: KDD 2018 (2018)
11. Zheng, G., et al.: A deep reinforcement learning framework for news recommendation. In: WWW 2018 (2018)
12. Zhu, Y.X., Lü, L.-Y.: Evaluation metrics for recommender systems. Dianzi Keji Daxue Xuebao/J. Univ. Electron. Sci. Technol. China **41**(2), 163–175 (2012)

13. Qin, L., Chen, S., Zhu, X.: Contextual combinatorial bandit and its application on diversified online recommendation. In: SDM 2014 (2014)

14. Mnih, V., et al.: 2015. Human-level control through deep reinforcement learning. Nature **518**, 7540, 529–533 (2015)

15. Wu, S., Ren, W., Yu, C., Chen, G.; Zhang, D., Zhu, J.: Personal recommendation using deep recurrent neural networks in NetEase. In: 2016 IEEE 32nd International Conference on Data Engineering (ICDE), pp. 1218–1229. IEEE (2016)

16. Liu, Y., Zhang, Y., Wu, Q., et al.: Diversity-promoting deep reinforcement learning for interactive recommendation. arXiv: Information Retrieval (2019)

17. Carbonell, J., Goldstein, J.: The use of MMR, diversity-based reranking for reordering documents and producing summaries. In: SIGIR 1998 (1998)

18. Lathia, N., Hailes, S., Capra, L., Amatriain, X.: Temporal diversity in recommender systems. In: SIGIR 2010 (2010)

19. Zhao, G., Lee, M.L., Hsu, W., Chen, W.: Increasing temporal diversity with purchase intervals. In: SIGIR 2012 (2012)

20. Qin, L., Zhu, X.: Promoting diversity in recommendation by entropy regularizer. In: IJCAI 2013 (2013)

21. Antikacioglu, A., Ravi, R.: Post processing recommender systems for diversity. In: KDD 2017 (2017)

22. Cheng, P., Wang, S., Ma, J., Sun, J., Xiong, H.: Learning to recommend accurate and diverse items. In: WWW 2017 (2017)

23. Chen, L., Zhang, G., Zhou, E.: Fast greedy map inference for determinantal point process to improve recommendation diversity. In: NIPS 2018 (2018)

Author Index

Printed in the United States
by Baker & Taylor Publisher Services